Table of Content

Table of Content	1
Introduction: Embrace the SEAL Mentality for Bugging In	23
Chapter 1: The Bug-In Essentials: Gear, Supplies, and Setup	25
1.1 - Understanding the Bug-In Strategy: Why Staying Put is Often the Smartest Option	25
1.2 - Performing a Threat Assessment: Tailoring Your Bug-In Strategy to Your Environment	26
The Bug-In Threat Matrix	27
1.3 - The Navy SEAL Gear Philosophy: Functionality, Versatility, Durability	28
Key Principles for Gear Selection:	28
1.4 - Building Your Bug-In Gear Arsenal: Must-Have Items	29
Water Filtration and Storage	29
First Aid Kits and Medical Supplies	30
Food Storage Solutions	31
Defense and Perimeter Security	32
Multitools and Tactical Gear	33
Communication and Power Solutions	33
1.5 - Establishing a Command Center: Organizing Your Supplies and Space	34
Choosing the Location	35
Stocking the Command Center	35
1.6 - Maintaining the SEAL Mindset During a Bug-In	36

Chapter 2: Securing Your Fortress: Perimeter Defense and Home Safety   **37**

  2.1 - Perimeter Defense: Establishing the First Line of Security   38
    Assessing Your Vulnerabilities   38
    Fencing and Barriers   39
    Lighting and Motion Detectors   40
    Creating a Buffer Zone   40

  2.2 - Reinforcing Entry Points: Doors, Windows, and Other Vulnerabilities   41
    Doors: Your First Line of Defense   41
    Windows: A Weak Spot if Left Unprotected   42
    Garage and Basement Doors: Hidden Vulnerabilities   43

  2.3 - Interior Defense: Creating Safe Rooms and Emergency Plans   44
    Safe Rooms: Your Last Line of Defense   44
    Defense Tools: Preparing for Close-Quarters Combat   45

  2.4 - Evacuation Planning: Knowing When to Bug Out   46
    Creating an Evacuation Plan   46

  2.5 - Psychological Barriers: Deterring Intruders Before They Act   47
    Signs and Warnings   47
    Dogs as Deterrents   48

  Conclusion: Preparing for the Worst, Securing the Best   48

Chapter 3: Mental Toughness: SEAL Psychology for Staying Calm in a Crisis   **49**

3.1 - The SEAL Mindset: How Mental Toughness is Built 50
    Commitment to Purpose 50
    The Power of Grit and Perseverance 52
    Stress Inoculation: Training the Brain to Handle Pressure 53

3.2 - Building Resilience: The SEAL Approach to Handling Setbacks 54
    Focus on What You Can Control 54
    The Power of Positivity and Visualization 55
    Embracing Hardship: Turning Discomfort Into Strength 56

3.3 - Maintaining Morale: Mental Health Strategies for Long-Term Isolation 57
    Creating Structure and Routine 58
    Managing Loneliness and Cabin Fever 59
    Team Dynamics: Managing Stress in Group Settings 60

Conclusion: The SEAL Blueprint for Mental Toughness 60

Chapter 4: Surviving Long-Term Power Outages: Adapt, Overcome, and Thrive 62

4.1 - Understanding the Impact of a Long-Term Power Outage 63
    Common Immediate Effects 63
    Developing a Power-Outage Survival Plan 64

4.2 - Lighting Solutions for a Grid-Down Scenario 65
    Battery-Powered Lanterns and Flashlights 65
    Solar-Powered and Crank Lanterns 66

Candles and Oil Lamps 66
Glow Sticks and Chemical Lights 67
4.3 - Temperature Control: Staying Warm or Cool Without Power 67
Staying Warm in Cold Climates 68
Portable Heaters and Fireplaces 69
Staying Cool in Hot Climates 70
4.4 - Ensuring a Steady Water Supply 71
Storing Water for Emergencies 71
Alternative Water Sources 72
Purifying Water 72
4.5 - Food Preservation and Cooking Without Power 73
Preserving Perishable Foods 73
Alternative Cooking Methods 74
4.6 - Staying Informed: Communication and Information in a Power Outage 75
Emergency Radios 75
Two-Way Radios 75
Staying Mentally Prepared 76
Conclusion: Adapting to Life Without Power 76

Chapter 5: Water Supply and Purification: Navy SEAL Hydration Tactics **77**
5.1 - Assessing Your Water Needs 77
How Much Water Do You Need? 78
5.2 - Storing Water for Long-Term Survival 79
Choosing Water Storage Containers 79
Sanitizing Water Storage Containers 80
Filling and Storing Water 80

Water Storage for Non-Drinkable Uses	81
5.3 - Alternative Water Sources During a Crisis	81
Rainwater Collection	82
Nearby Natural Water Sources	82
Well Water	83
5.4 - Purifying Water for Safe Drinking	83
Boiling Water	84
Chemical Water Purification	84
Water Filtration	85
Solar Water Disinfection (SODIS)	86
5.5 - Maintaining Hygiene and Sanitation with Limited Water	86
Conserving Water for Hygiene	87
Sanitation and Waste Management	87
5.6 - Final Thoughts: Water is Life	88
Chapter 6: Food Storage for Extended Bug-Ins: What SEALs Eat to Survive	89
6.1 - Assessing Your Food Needs: How Much Should You Store?	90
Daily Caloric Requirements	90
Calculating Total Food Storage Needs	91
6.2 - Types of Food to Store: Building a Balanced Stockpile	91
Non-Perishable Foods for Long-Term Storage	92
Freeze-Dried and Dehydrated Foods	95
MREs (Meals Ready to Eat)	96
6.3 - Food Preservation Methods: Extending Shelf Life	96

Canning and Jarring	97
Dehydration	97
Fermentation	98
Salting and Smoking	99
6.4 - Meal Planning and Rationing	99
Meal Planning for Efficiency	99
Rationing Your Food Supply	100
6.5 - Cooking Without Electricity	101
Propane and Butane Stoves	101
Wood-Burning Stoves and Fireplaces	101
Solar Cookers	101
6.6 - Final Thoughts: Nutrition and Survival	102
Chapter 7: Medical Preparedness: Self-Care and First Aid Training for Emergencies	104
7.1 - Building a Comprehensive First Aid Kit: Essential Medical Supplies	105
Basic First Aid Supplies	105
Advanced First Aid Supplies	107
Medications and Personal Health Supplies	109
7.2 - First Aid Skills: Basic Training Everyone Should Know	110
How to Perform CPR	110
How to Control Bleeding	111
How to Treat Shock	111
How to Splint a Broken Bone	112
7.3 - Medical Conditions to Be Aware Of: Illnesses and Infections	113
Dehydration	113

    Infections and Sepsis   114
    Heat Exhaustion and Heatstroke   115
    Hypothermia   115
7.4 - Creating a Medical Emergency Plan   116
7.5 - Final Thoughts: Becoming Your Own First Responder   117

Chapter 8: Communication and Intelligence: Staying Informed and Connected   118

8.1 - Why Communication Is Critical in a Crisis   119

8.2 - Emergency Communication Methods: Staying Connected Without the Grid   120
    AM/FM and NOAA Radios   120
    Hand-Crank and Solar-Powered Radios   120
    Two-Way Radios (Walkie-Talkies)   121
    CB Radios (Citizens Band Radios)   122
    Ham Radios (Amateur Radios)   123
    Satellite Phones   123

8.3 - Communication Security: Protecting Your Information   124
    Encryption and Secure Channels   124
    Operational Security (OPSEC)   125

8.4 - Gathering Intelligence: Staying Informed During a Crisis   126
    Monitoring Local News and Emergency Broadcasts   126
    Intelligence from Neighbors and Local Networks   127
    Online Sources and Social Media   128

8.5 - Psychological Preparedness: Staying Calm

in Isolation   129
- Combatting Isolation and Loneliness   129
- Stress Management and Mental Resilience 129

8.6 - Final Thoughts: Communication as a Lifeline   130

Chapter 9: Defending Your Home: Security Tactics for Bugging In   132

9.1 - Establishing Layers of Defense: Creating a Secure Perimeter   133
- The First Layer: Perimeter Security   133
  - Fencing and Gates   133
  - Lighting and Cameras   134
- The Second Layer: Yard or Entryway Defense   135
  - Defensive Landscaping   135
  - Obstacles and Barriers   136

9.2 - Reinforcing Entry Points: Securing Doors, Windows, and Access Points   137
- Reinforcing Doors   137
  - Solid-Core or Metal Doors   137
  - Security Doors   138
- Securing Windows   138
  - Window Locks and Bars   138
  - Reinforced Glass and Window Film   139
- Securing the Garage   140
  - Garage Door Locks and Bars   140
  - Disable the Emergency Release   140

9.3 - Tactical Home Defense: Preparing for Worst-Case Scenarios   141

    Safe Rooms and Retreat Points   141
    Self-Defense Tools   142
    Emergency Drills   142

9.4 - Psychological Preparedness: Staying Calm Under Threat   143
    Dealing with Fear   143
    Situational Awareness   143

9.5 - Final Thoughts: Securing Your Fortress   144

Chapter 10: Sustainable Living: Creating Long-Term Systems for Water, Food, and Energy   146

10.1 - Establishing a Long-Term Water System   146
    Rainwater Harvesting   146
        Setting Up a Rainwater Collection System   147
        Maximizing Water Storage   148
    Alternative Water Sources   148

10.2 - Sustainable Food Production: Growing and Preserving Your Own Food   149
    Creating a Survival Garden   149
        Planning Your Garden   150
        Extending Your Growing Season   151
    Raising Livestock and Poultry   151
        Chickens for Eggs and Meat   152
        Goats for Milk and Meat   152
    Preserving Food for Long-Term Storage   153
        Canning   153
        Dehydration   154
        Fermentation   154

- 10.3 - Sustainable Energy Solutions: Powering Your Home Without the Grid ... 155
  - Solar Power ... 155
    - Solar Panels ... 155
    - Battery Storage ... 156
  - Generators ... 156
    - Gas Generators ... 157
    - Propane and Solar Generators ... 157
- 10.4 - Waste Management and Sanitation Systems ... 158
  - Composting Toilets ... 158
  - Graywater Systems ... 159
- 10.5 - Psychological Resilience: Adapting to Sustainable Living ... 160
  - Adapting to a New Routine ... 160
  - Staying Motivated ... 160
  - Building Community Support ... 160
- 10.6 - Final Thoughts: Thriving Through Sustainability ... 161

Chapter 11: Mental and Emotional Resilience: Maintaining Psychological Strength in Long-Term Crises ... 163

- 11.1 - Understanding Psychological Resilience: What Navy SEALs Teach Us ... 164
  - The Importance of Mental Toughness ... 164
  - The Stress Response: Understanding Fight, Flight, and Freeze ... 165
- 11.2 - Managing Stress: SEAL Techniques for Staying Calm and Focused ... 165
  - Controlled Breathing (Box Breathing) ... 165

Visualization and Mental Rehearsal 166
Emotional Control: Responding, Not Reacting 167
11.3 - Overcoming Isolation and Loneliness: Maintaining Social Connections 168
Staying Connected: Even When Physically Isolated 169
Creating a Routine and Structure 169
Mental and Emotional Engagement 170
11.4 - Handling Crisis Fatigue: Dealing with Prolonged Stress 171
Recognizing the Signs of Crisis Fatigue 172
Creating Time for Recovery 172
Setting Boundaries and Managing Expectations 173
11.5 - Cultivating Hope and Positivity in Dark Times 174
The Power of a Positive Mindset 174
Building a Sense of Purpose 175
11.6 - Final Thoughts: The Strength of the Human Spirit 176

Chapter 12: Managing Medical Emergencies: Field Medicine and Healthcare Preparedness 177
12.1 - Building a Comprehensive Medical Kit: Essential Supplies for Every Scenario 178
Basic First Aid Supplies 178
Advanced First Aid and Trauma Supplies 179
Medications and Preventive Supplies 181
12.2 - Managing Medical Emergencies: Essential Skills for Every Household 182

How to Perform CPR 182
How to Control Severe Bleeding 183
How to Treat Burns 184
How to Splint a Broken Bone 184
How to Treat Shock 185

12.3 - Treating Common Illnesses and Infections: Maintaining Health During a Crisis 186

Dehydration 186
Infections and Sepsis 187
Gastrointestinal Illnesses 187
Heat Exhaustion and Heatstroke 188

12.4 - Preventive Care and Hygiene: Keeping Your Household Healthy 189

Personal Hygiene Practices 189
Sanitizing Your Environment 190
Boosting Immune Health 190

12.5 - Final Thoughts: Medical Preparedness as a Lifeline 191

Chapter 13: Long-Term Security and Protection: Evolving Your Defenses for Sustained Crisis Survival **193**

13.1 - Layered Defense Systems: Strengthening and Adapting Your Perimeter 194

Reinforcing Your Perimeter: Evolving Threat Response 194

Barriers and Deterrents 194
Improving Visibility and Surveillance 195

Managing Stealth: Avoiding Unwanted Attention 196

Reducing Noise and Light Pollution 196

Concealing Your Supplies and Activity 197
13.2 - Contingency Plans: Preparing for a Breach 198
- Safe Rooms and Retreat Areas 198
  - Choosing a Safe Room 198
  - Stocking Your Safe Room 198
- Home Defense and Self-Defense Tactics 199
  - Firearms for Home Defense 199
  - Non-Lethal Self-Defense Options 200
  - Using the Environment to Your Advantage 201

13.3 - Handling Confrontations: SEAL Strategies for De-Escalation and Engagement 202
- De-Escalating Threats: When Talking is the Best Strategy 202
  - Body Language and Non-Verbal Cues 202
  - Verbal Communication 203
- When Engagement is Necessary: SEAL Combat Principles 203
  - Positioning and Tactical Advantage 203
  - Commit to Action 204

13.4 - Adapting to Changing Threats: Long-Term Security Considerations 205
- Threat Assessment and Intelligence Gathering 205
- Managing Resource Security 205

13.5 - Final Thoughts: The Evolution of Home Security 206

Chapter 14: Building Community Networks: Leveraging Cooperation for Long-Term Survival 208

14.1 - The Importance of Community in a Long-Term Crisis — 208
    Why Community Matters in a Crisis — 208
    Balancing Community and Self-Reliance — 209

14.2 - Forming Alliances: Building Trust with Neighbors — 210
    Identifying Potential Allies — 210
    Building Trust in a Crisis — 211
        Start with Small Gestures — 211
        Demonstrate Competence and Reliability — 211
        Develop Clear Communication Channels — 212
    Forming a Community Watch Group — 213

14.3 - Organizing Roles and Responsibilities: A Team Approach to Survival — 213
    Assigning Roles Based on Skills — 213
    Rotating Responsibilities to Prevent Burnout — 215
    Planning for the Long-Term: Sustainable Roles — 215

14.4 - Sharing Resources and Bartering: Managing Community Supplies — 216
    Establishing a Community Resource Pool — 216
        What to Pool — 216
        Rationing and Distribution — 217
    Bartering Within the Community — 218
    Managing Conflicts Over Resources — 218

14.5 - Handling Community Challenges: Conflict Resolution and Leadership — 219

The Role of Leadership 219
    Choosing a Leader 220
    Rotating Leadership Roles 220
Managing Community Conflicts 221
    Common Sources of Conflict 221
    Conflict Resolution Strategies 221
14.6 - Final Thoughts: The Strength of a United Community 222

Chapter 15: Psychological Warfare: Defending Against Manipulation, Coercion, and Intimidation 224

15.1 - Understanding Psychological Warfare in a Crisis 225
    Why Psychological Warfare Happens in Crises 225
    Forms of Psychological Warfare 225
15.2 - Recognizing Psychological Manipulation and Intimidation Tactics 227
    Spotting Intimidation Tactics 227
    Recognizing Manipulation Tactics 228
    Understanding "Divide and Conquer" Tactics 229
15.3 - Building Mental Defenses: Resisting Psychological Manipulation and Coercion 230
    Mental Toughness and Confidence 230
        Breathing Techniques and Mindfulness 230
        Positive Visualization 231
    Critical Thinking and Skepticism 231
        Question Everything 231
        The Power of "No" 232

15.4 - Countering Psychological Warfare: SEAL Strategies for Turning the Tables 233
   Control the Narrative: Shaping Perception 233
      Appear Strong, Even When You're Not 233
      Misinformation as a Defense 234
   Deflecting and Reversing Manipulation 234
      Turn the Focus Back on the Manipulator 235
      Offer an Unattractive Alternative 235
15.5 - Protecting Your Community from Psychological Warfare 236
   Establish Clear Communication Channels 236
   Foster Unity and Trust 237
   Neutralizing External Threats 237
15.6 - Final Thoughts: Fortifying Your Mind for Survival 238

Chapter 16: Long-Term Sustainability: Adapting Your Home and Lifestyle for Indefinite Survival 240
16.1 - Transitioning from Short-Term to Long-Term Survival 240
   Recognizing When to Transition to Long-Term Planning 241
      Signs It's Time to Shift Your Focus: 241
   Mindset Shift: From Survival to Thriving 242
16.2 - Sustainable Food Production: Growing and Preserving Your Own Food 242
   Starting a Survival Garden 242
      Choosing What to Grow 243

Maximizing Space	244
Growing Year-Round	244
Preserving Food for Long-Term Storage	244
Canning	245
Dehydration	245
Fermentation	246
Freezing	246
16.3 - Sustainable Water Management: Ensuring a Continuous Supply of Clean Water	247
Rainwater Harvesting Systems	247
Setting Up a Rainwater Collection System	247
Filtering and Purifying Rainwater	248
Alternative Water Sources	248
16.4 - Energy Independence: Powering Your Home Without the Grid	249
Solar Power: The Best Long-Term Energy Solution	250
Setting Up Solar Panels	250
Powering Essential Devices	250
Backup Generators: A Temporary Power Solution	251
Types of Generators	251
Managing Fuel and Power Usage	252
16.5 - Mental and Emotional Resilience: Thriving Through Psychological Endurance	253
Maintaining Routine and Structure	253
Creating a Daily Schedule	253
Building Community Support	254

- Staying Motivated and Setting Goals — 254
- Managing Stress and Anxiety — 255
- 16.6 - Final Thoughts: Thriving Through Sustainability and Adaptability — 256

Chapter 17: Adapting to Changing Threats and Environmental Conditions — 257

- 17.1 - Identifying and Understanding New Threats — 257
  - Assessing Evolving Security Risks — 257
    - Signs of Escalating Security Threats — 258
    - Adjusting Your Security Strategy — 259
  - Monitoring New Environmental Challenges — 260
    - Identifying Environmental Changes — 260
    - Adapting to Environmental Threats — 261
- 17.2 - Resource Scarcity: Adapting Your Supply Management — 262
  - Conserving Supplies and Reducing Waste — 262
    - Food and Water Conservation — 262
    - Energy and Fuel Conservation — 263
  - Securing Alternative Supply Sources — 263
    - Foraging and Hunting — 264
    - Bartering and Trade — 264
- 17.3 - Flexibility in Shelter: Adjusting Your Living Environment — 265
  - Expanding or Modifying Your Shelter — 265
    - Adding Outdoor Structures — 266
    - Reinforcing Your Home's Defenses — 267
  - Relocating or Retreating — 267

Bug-Out Locations	267
Go-Bags	268
17.4 - Adapting Your Mindset: Embracing Change and Overcoming Fatigue	268
Dealing with Fatigue and Burnout	268
Building Recovery Periods into Your Routine	269
Flexibility in Decision-Making	269
Adjusting Strategies	270
Learning from Mistakes	270
17.5 - Final Thoughts: Flexibility as a Survival Tool	271
Chapter 18: Strategic Leadership and Decision-Making in a Long-Term Crisis	**272**
18.1 - The Role of a Leader in a Survival Scenario	273
Key Responsibilities of a Leader	273
Traits of an Effective Crisis Leader	273
18.2 - Decision-Making Under Pressure	274
The OODA Loop: A Framework for Decision-Making	275
Balancing Risk and Reward	276
Trusting Your Instincts	276
18.3 - Maintaining Group Morale and Unity	277
The Importance of Purpose	277
Creating a Shared Goal	278
Managing Stress and Anxiety	278
Signs of Stress and Burnout	279
Promoting Rest and Recovery	279

Conflict Resolution and Group Cohesion 280
   Addressing Conflict Early 280
   Building Unity Through Shared Experiences 281
18.4 - Navigating Leadership Challenges in a Crisis 282
   Balancing Individual and Group Needs 282
      Setting Group Priorities 282
   Handling Difficult Decisions 283
      Clear Communication in Decision-Making 283
   Managing Leadership Fatigue 283
      Delegating Responsibility 284
      Taking Breaks 284
18.5 - Final Thoughts: Leadership as a Lifeline in Long-Term Survival 285

Chapter 19: Advanced Medical Care and First Aid: Self-Reliance in Long-Term Survival 286
   19.1 - Building an Advanced Medical Kit for Long-Term Survival 286
      Basic First Aid Supplies 286
      Advanced Medical Supplies 287
      Long-Term Health Maintenance Supplies 289
   19.2 - Treating Common Injuries in a Survival Scenario 290
      Wound Care and Infection Prevention 290
         Step-by-Step Wound Treatment 290
      Burns 291
         Step-by-Step Burn Treatment 292
      Sprains and Fractures 292

Treating a Sprain	293
Treating a Fracture	293
19.3 - Managing Chronic Illnesses in a Long-Term Crisis	294
Diabetes Management	294
Stockpiling Insulin and Supplies	294
Dietary Adjustments	295
Asthma and Respiratory Conditions	295
Stockpiling Inhalers and Medication	295
Improving Air Quality	296
Heart Disease and High Blood Pressure	296
Medication Management	296
Lifestyle Adjustments	297
19.4 - Emergency Trauma Care and Triage	297
Performing Triage in a Survival Situation	298
Triage Categories	298
Emergency Trauma Care Techniques	298
Managing Severe Bleeding	299
Treating Head and Neck Injuries	299
Treating Chest Trauma	300
Treating Shock	301
19.5 - Final Thoughts: Building Confidence in Your Medical Skills	301
Chapter 20: Mental and Emotional Resilience: Thriving in a Long-Term Crisis	303
20.1 - Understanding the Psychological Toll of a Long-Term Crisis	303
Common Psychological Challenges in Long-Term Crises	304

- Anticipating the Psychological Effects — 305
- 20.2 - Building Mental Resilience: The Navy SEAL Mindset — 305
  - Components of Mental Resilience — 305
  - Mental Toughness Exercises — 307
- 20.3 - Creating Structure and Routine for Psychological Stability — 308
  - Establishing a Daily Routine — 308
    - Productive Tasks — 309
    - Physical Activity and Exercise — 309
    - Rest and Relaxation — 310
  - The Power of Routine in Group Dynamics — 310
- 20.4 - Coping with Isolation and Loneliness — 311
  - The Effects of Isolation — 311
  - Combatting Loneliness and Maintaining Social Connections — 312
    - Strengthen Relationships Within Your Household — 312
    - Stay Connected Remotely — 312
    - Create New Social Rituals — 312
- 20.5 - Maintaining Hope and Motivation in Long-Term Survival — 313
  - The Importance of Hope — 313
    - Celebrate Small Victories — 313
    - Set Short-Term Goals — 314
    - Maintain a Positive Outlook — 314
    - Inspire Others — 314
- 20.6 - Final Thoughts: Surviving and Thriving Through Mental Strength — 315

Introduction: Embrace the SEAL Mentality for Bugging In

When the world outside turns chaotic, whether due to natural disasters, social unrest, or a sudden crisis, bugging in—hunkering down in your own home—becomes not just a smart option, but a necessary one. The **Navy SEALs Bug-In Guide** is designed to teach you how to navigate this challenge with the precision and preparedness of one of the world's most elite special operations forces.

This guide isn't just about stockpiling supplies and hoping for the best. It's about adopting a **warrior's mindset** and **methodical approach** to long-term survival while staying put in your home. As a Navy SEAL, you're trained to adapt to harsh environments, think on your feet, and make life-or-death decisions under immense pressure. The same principles can be applied to a bug-in situation.

You don't need to live in a bunker or be armed to the teeth to benefit from these tactics. You can turn your home into a fortress of security, stability, and comfort with the right knowledge and mindset. As you work through this guide, you'll realize that true preparedness comes not just from having the right tools but from adopting the right mentality—the SEAL mentality.

A SEAL knows that **preparedness is not paranoia**—it's prudence. This book will show you how to fortify your home, make it self-sustaining, and protect your family using strategies honed on battlefields and in some of the world's most dangerous environments. You'll learn how to outthink threats, adapt to changing circumstances, and maintain your composure no matter what's happening outside.

In the coming chapters, we will dive into **tactical preparedness** from the perspective of a Navy SEAL: securing your space, maintaining supplies, and protecting your loved ones. Whether it's a natural disaster, a pandemic, or civil unrest, the methods described here will equip you with the mental and physical strategies needed to outlast the storm—without ever having to leave your home.

Chapter 1: The Bug-In Essentials: Gear, Supplies, and Setup

1.1 - Understanding the Bug-In Strategy: Why Staying Put is Often the Smartest Option

In times of crisis—whether it's a natural disaster, pandemic, societal breakdown, or economic collapse—your home can become your strongest asset. While many people might think fleeing (or "bugging out") to a remote location is the best option, the reality is that **staying home and fortifying your residence** is often a far more viable strategy. This decision is known as "bugging in."

The key advantage of a bug-in strategy is familiarity. You know your home better than any place in the world. You know its strengths, its vulnerabilities, its hiding spots, and how to navigate it in the dark. Plus, you're already equipped with many of the supplies you'll need to sustain yourself for an extended period of time. But the success of a bug-in doesn't rely solely on being in your home; it requires **preparation, strategic planning**, and, most importantly, a **mental shift** toward self-sufficiency.

Navy SEALs are trained to adapt to extreme conditions and to survive in hostile environments. The **SEAL mindset** is built on principles of readiness, resilience, and resourcefulness, all of which are essential during a bug-in scenario. In this chapter, we'll explore how to prepare your home for an extended crisis, what gear you need to gather, and how to think like a Navy SEAL when planning for the unexpected.

1.2 - Performing a Threat Assessment: Tailoring Your Bug-In Strategy to Your Environment

Before diving into specific gear and supplies, the first step in a successful bug-in is conducting a **threat assessment**. Every environment presents its own unique risks, and understanding these threats allows you to tailor your preparations effectively. Navy SEALs always conduct thorough threat assessments before every mission, and you should do the same for your home.

Start by asking yourself the following questions:

- What natural disasters are most likely to occur in your region? (e.g., hurricanes, earthquakes, tornadoes, wildfires)
- What are the chances of a power grid failure in your area?
- How vulnerable is your neighborhood to social unrest or looting?

- How dependent are you on external resources like water, electricity, and food?

Once you've identified the most probable threats, you can begin creating a **Bug-In Threat Matrix**, which helps you rank the risks based on likelihood and severity. This matrix will become your roadmap for prioritizing which supplies and preparations are most critical. For example, if you live in an earthquake-prone area, structural reinforcement and emergency medical supplies might take precedence. If you're in a region vulnerable to hurricanes, your focus may be more on securing windows, stockpiling water, and preparing for long-term power outages.

The Bug-In Threat Matrix

Your Bug-In Threat Matrix should look something like this:

Threat	Likelihood (High/Medium/Low)	Impact (Severe/Moderate/Minimal)	Preparations
Earthquake	High	Severe	Structural reinforcement, first aid
Hurricane	Medium	Severe	Water, non-perishables, generators
Power grid failure	Low	Moderate	Solar panels, backup batteries
Social unrest/looting	Low	Severe	Perimeter security, defense gear
Pandemic	Medium	Moderate	PPE, hygiene supplies, isolation plans

By filling out this matrix, you'll get a clearer sense of the **immediate vulnerabilities** your home and family face, which in turn informs the types of **gear and supplies** you'll need to prioritize.

1.3 - The Navy SEAL Gear Philosophy: Functionality, Versatility, Durability

Navy SEALs operate in some of the most hostile environments in the world, and they depend on their gear for survival. When selecting gear for your bug-in strategy, you need to adopt the same philosophy that SEALs use: **gear should be functional, versatile, and durable**.

SEALs are trained to carry only what is necessary for the mission, and each piece of equipment must serve multiple purposes. When preparing for a bug-in, the idea isn't to amass as many items as possible; rather, it's to collect **critical items that can serve you in multiple ways** and ensure that these items are of high quality and long-lasting.

Key Principles for Gear Selection:

- **Functionality**: Every item should have a clear and specific function. Avoid "gadgets" that have little real-world application.

- **Versatility**: Can one item serve multiple purposes? For example, a **multitool** can be used for cutting, screwing, and repairing, which makes it more valuable than a single-use tool.
- **Durability**: Your gear needs to last, especially in high-stress environments. Cheap gear might save money upfront, but in a crisis, durability can mean the difference between life and death.

1.4 - Building Your Bug-In Gear Arsenal: Must-Have Items

Let's break down the essential categories of gear you'll need during a bug-in, along with the specific items recommended based on **Navy SEALs' survival training**.

Water Filtration and Storage

Water is the most important resource for survival. While SEALs are trained to find water sources in the wild, in a bug-in scenario, your home's water supply could become contaminated or shut off completely. You'll need a combination of **stored water** and **water filtration systems** to ensure your family has a clean and safe supply throughout the crisis.

- **Stored Water**: Aim to have at least **one gallon per person, per day** for drinking, cooking, and hygiene. A two-week supply is the minimum recommended, but for extended bug-ins,

consider storing **large water barrels** with purification tablets.
- **Water Filtration Systems**: In case your water supply runs out, you'll need a way to purify water from alternative sources like lakes or rainwater. The **LifeStraw** or **Sawyer Mini** are compact, portable water filters that can purify up to 100,000 liters of water. For larger-scale filtration, consider a **gravity-fed water filter** like the Berkey system, which can handle a household's water needs.

First Aid Kits and Medical Supplies

Medical emergencies can arise at any time, especially during a prolonged bug-in when professional help may not be immediately available. SEALs are trained in **field medicine**, and having the right medical supplies can help you treat injuries, infections, and other health concerns until medical assistance is available.

- **Basic First Aid Kit**: A well-stocked first aid kit should include **bandages, gauze, antiseptics, burn cream, splints, pain relievers, antihistamines, and antibiotics** if possible.
- **Specialized Medical Supplies**: Depending on your family's needs, you may want to include supplies for treating specific conditions like asthma (inhalers), diabetes (insulin), or severe allergies (epinephrine injectors).

- **Training**: Knowing how to use the supplies in your kit is as important as having them. Take a basic **first aid course** or study resources that teach you how to treat wounds, burns, fractures, and other common injuries.

Food Storage Solutions

Food is another vital component of your bug-in arsenal, and your goal should be to stockpile **non-perishable, calorie-dense foods** that can sustain your family for at least a month. SEALs are trained to survive on minimal rations, but during a bug-in, you want to maintain energy levels and morale by ensuring your family has adequate nourishment.

- **Canned Goods**: Stock up on **high-protein canned foods** like beans, fish (e.g., tuna, sardines), and meats (e.g., canned chicken or beef). Also, include **canned vegetables and fruits** to maintain a balanced diet.
- **Freeze-Dried Meals**: Brands like **Mountain House** and **Wise Company** offer freeze-dried meals that have long shelf lives (up to 25 years) and only require water to prepare. These are ideal for bug-ins since they're lightweight and easy to store.
- **Staples and Energy-Dense Foods**: Items like **rice, pasta, peanut butter, oats**, and **nuts** provide long-lasting, energy-dense options that can form the basis of meals.

- **Comfort Foods**: Including some comfort foods like **chocolate, coffee, tea, and instant soups** can help maintain morale during a long bug-in.

Defense and Perimeter Security

Protecting your home is crucial during a bug-in. As SEALs know, security begins with **early detection** and **prevention**. You want to make your home as unattractive a target as possible while also being prepared to defend it if necessary.

- **Fortifying Your Home**: Reinforce doors and windows with **high-quality deadbolts, window bars, and shatterproof glass film**. If looting or home invasions are a concern, consider installing **metal or wood reinforcements** behind your doors.
- **Perimeter Security**: Install **motion-activated lights** and **security cameras** around your property to detect potential threats early. Outdoor lights act as a deterrent to intruders, and security cameras can alert you to movement near your home.
- **Defense Tools**: While some individuals may choose firearms for home defense, other non-lethal options include **pepper spray, tasers**, or **baseball bats**. SEALs are trained to use whatever is at hand to defend themselves, and you should be equally adaptable.

Multitools and Tactical Gear

SEALs rely heavily on their tools, and a good **multitool** can be your best friend in a bug-in scenario. These versatile devices offer a wide range of functions that can help with repairs, opening cans, cutting materials, and more.

- **Multitools**: A high-quality multitool, such as a **Leatherman** or **Swiss Army Knife**, should be part of your daily carry during a bug-in. Look for tools that include **pliers, wire cutters, knives, screwdrivers, bottle openers**, and **scissors**.
- **Tactical Flashlights**: A durable, tactical flashlight is essential for any power outage scenario. Choose a **LED flashlight** with at least **1000 lumens** for bright, far-reaching light. Some tactical flashlights also come with **strobe functions** for disorienting intruders.
- **Solar-Powered Chargers**: You'll need a way to charge essential electronics during a prolonged power outage. Solar-powered chargers are compact, efficient, and can keep your **phones, radios, and other small devices** operational.

Communication and Power Solutions

During a bug-in, staying informed and connected to the outside world is critical. SEALs rely on **radios** and **satellite communications** to receive mission updates, and you can use similar tools to keep track of what's happening in your area.

- **Hand-Crank or Solar-Powered Radios**: These radios can pick up **AM/FM and NOAA weather channels**, providing you with important updates on the status of the crisis. They often come equipped with hand-crank or solar charging capabilities, ensuring you don't need external power sources.
- **Two-Way Radios**: In case of a total communication breakdown (like a cell network failure), having **two-way radios** will allow you to stay in contact with family members both inside and outside your home.
- **Backup Power**: Investing in a **generator** or **solar power system** can provide you with electricity during extended outages. Generators should be used outdoors and far from windows to avoid carbon monoxide poisoning, while solar power systems are quieter and eco-friendly.

1.5 - Establishing a Command Center: Organizing Your Supplies and Space

Just as Navy SEALs have a central command center during operations, you'll want to designate a specific area of your home as your **Command Center** during a bug-in. This is the place where you'll store and organize all your critical gear, supplies, and information.

Choosing the Location

Your command center should be:

- **Secure**: Choose a room without many windows and with a solid door that can be reinforced if necessary.
- **Accessible**: Make sure that the location is easy to access from other parts of the house, especially in an emergency.
- **Organized**: Use **shelves, containers, and labels** to keep everything organized and easy to find. Your command center should be laid out in a way that allows you to quickly grab what you need without wasting time searching.

Stocking the Command Center

In your command center, you should store:

- **First aid supplies**
- **Communication devices (radios, backup phones)**
- **Tactical gear and defense tools**
- **Maps, threat assessments, and family emergency plans**
- **Extra water and food supplies**

By centralizing your most important resources, you'll ensure that you have a go-to location for managing your bug-in. This level of **organization and preparedness** is what separates a successful bug-in from a chaotic one.

1.6 - Maintaining the SEAL Mindset During a Bug-In

Beyond the gear and supplies, the most critical aspect of a successful bug-in is **your mindset**. Navy SEALs are trained to stay calm under pressure, adapt to changing situations, and maintain mental and physical discipline. The same principles apply to bugging in. You must be ready to **embrace the uncertainty** of a prolonged crisis and maintain the ability to think critically in high-stress situations.

- **Stay Calm**: Stress can cloud your judgment. Practice **stress-management techniques** like deep breathing, meditation, or even SEAL breathing exercises to keep a clear head.
- **Adaptability**: Flexibility is key. If circumstances change, you may need to alter your bug-in plan on the fly. A SEAL knows that no plan survives first contact with the enemy, and you must be willing to adjust.
- **Teamwork and Communication**: If you're bugging in with family members, clear **communication and teamwork** are essential. Assign roles to each person and practice working together, just as SEAL teams do.

In conclusion, **preparedness is power**. With the right gear, a fortified home, and the SEAL mindset, you'll be ready to face whatever challenges a bug-in scenario throws your way.

Chapter 2: Securing Your Fortress: Perimeter Defense and Home Safety

In a true crisis scenario, whether due to civil unrest, natural disasters, or any other external threat, your home becomes your castle. Just as Navy SEALs rely on fortified bases and strongholds to survive in hostile environments, you too must transform your home into a **fortress**—a place of safety that can withstand external threats while keeping those inside protected. This chapter is focused on how to reinforce your home, create layers of security, and develop both physical and psychological barriers that deter intruders or threats.

When we talk about security, it isn't just about installing a deadbolt or buying a firearm. It's about adopting a **holistic approach** that encompasses everything from **perimeter security**, **entry point reinforcement**, **defense planning**, and **emergency evacuation strategies**. A SEAL's approach to any mission involves thorough preparation, situational awareness, and redundancy, and these principles apply directly to home defense during a bug-in scenario.

2.1 - Perimeter Defense: Establishing the First Line of Security

Your first line of defense is the **perimeter** around your property. Whether you live in a house with a yard or a small apartment, thinking about the perimeter is critical to keeping threats at bay. The perimeter includes everything from the outer edges of your property (fences, gates, walls) to the area directly surrounding your home (doors, windows, outdoor spaces).

Assessing Your Vulnerabilities

The first step is to walk the perimeter of your property and **assess vulnerabilities**. Take note of weak spots where an intruder could enter, either by force or by stealth. Think like an enemy: If you wanted to break into your own home, where would you start? Common vulnerabilities include:

- **Fences** that are too short or easy to scale
- **Bushes or trees** that provide cover for someone trying to approach the house unnoticed
- **Unlit areas** that create shadows and blind spots
- **Windows** on the ground floor that are easily accessible or poorly secured
- **Garage doors** that might not be reinforced

Navy SEALs are trained to exploit the weaknesses of their enemies, but in this case, your goal is to identify your weaknesses before someone else can. Once you've identified these areas, you can begin fortifying them.

Fencing and Barriers

A **strong fence** can be your first barrier against unwanted intrusions. While a standard wooden or chain-link fence offers minimal protection, there are ways to reinforce these structures to provide additional security:

- **Height**: The taller the fence, the harder it is to scale. Aim for a fence that's at least **6-8 feet high**, which will deter casual intruders. For added security, consider topping your fence with **barbed wire** or **thorny plants**.
- **Visibility**: While it may seem counterintuitive, a fence that provides **visibility** (such as wrought iron or chain-link) can sometimes be more secure than a solid wood or brick wall. Solid walls provide cover for intruders, allowing them to hide while they work on gaining entry. A visible fence, on the other hand, offers no such cover.
- **Gates**: Any gate should be as secure as the rest of the fence. Use **heavy-duty locks** and ensure the gate cannot be easily removed from its hinges. SEALs know that even the smallest weakness can compromise a secure area, so treat your gates with the same seriousness you would a door.

Lighting and Motion Detectors

One of the simplest and most effective ways to secure your perimeter is with **outdoor lighting**. Darkness provides cover for intruders, and well-placed lights can remove that advantage:

- **Motion-Activated Lights**: These are a must-have for any bug-in scenario. Place them around **entry points**, **gates**, **garage doors**, and **dark corners** of your property. Motion-activated lights not only startle potential intruders, but they also alert you to their presence.
- **Solar-Powered Lights**: In the event of a power outage, solar-powered outdoor lights provide an additional layer of security. These lights store energy during the day and automatically activate at night.
- **Floodlights**: For large properties, **floodlights** can be installed to cover wide areas, such as driveways or backyards. These lights are typically placed higher up and are more powerful than standard motion lights.

Creating a Buffer Zone

A **buffer zone** is the area between your perimeter and your home. In military operations, this zone is critical for monitoring enemy movement and preventing them from approaching unnoticed. You can create a buffer zone around your home by:

- **Clearing vegetation**: Cut back any bushes, trees, or shrubs that are close to your home. These can provide cover for someone trying to sneak up on your house.
- **Gravel paths**: Consider laying **gravel** around the perimeter of your house. Gravel makes noise when walked on, alerting you to any movement outside.
- **Strategic planting**: Use **thorny bushes** (like holly or rose bushes) beneath windows or along the perimeter to create natural barriers that are difficult to penetrate.

2.2 - Reinforcing Entry Points: Doors, Windows, and Other Vulnerabilities

Once an intruder reaches your home, the next target is the **entry points**—doors, windows, and any other openings like garages or vents. Each of these entry points needs to be fortified to withstand both **forced entry** and **sneak attacks**.

Doors: Your First Line of Defense

The doors of your home are the most obvious entry points and should be fortified accordingly. A standard door with a basic lock is not sufficient to withstand a determined intruder. Here's how to upgrade your door defenses:

- **Solid Core Doors**: Ensure that all exterior doors are **solid core**, meaning they are made from wood or metal rather than hollow-core. Hollow-core doors are easily kicked in.
- **Deadbolt Locks**: Install **heavy-duty deadbolts** on all exterior doors. The bolt should extend at least **1 inch into the doorframe**. Opt for a **double-cylinder deadbolt**, which requires a key to open from both sides, preventing someone from breaking a window to reach inside and unlock the door.
- **Reinforced Door Frames**: Even the strongest door is only as good as the frame that supports it. Reinforce your door frames with **metal strike plates** and **long screws** that anchor deeply into the studs of the wall, making it much harder to kick in.
- **Security Bars**: For added protection, consider using a **security bar** or **door brace** on the inside of the door. These devices create additional resistance to forced entry.

Windows: A Weak Spot if Left Unprotected

Windows are one of the most vulnerable parts of your home, especially on the ground floor. To make them more secure, you need to approach window defense from both **inside and outside** the home:

- **Window Locks**: All windows should be equipped with **window locks**, even if they are rarely

opened. Simple latch locks are not enough; invest in **keyed window locks** that prevent windows from being opened without a key.
- **Shatterproof Film**: Applying **shatterproof film** to your windows makes them much harder to break. This film holds the glass together even when shattered, which slows down intruders and creates more noise, giving you time to react.
- **Window Bars and Grills**: If you live in an area prone to break-ins or expect significant social unrest, consider installing **window bars** or **security grills** on ground-floor windows. These bars can be designed to look aesthetically pleasing while still providing a strong barrier against entry.
- **Window Alarms**: Small, **battery-operated window alarms** can alert you if a window is opened or broken. These alarms are inexpensive and easy to install.

Garage and Basement Doors: Hidden Vulnerabilities

Often overlooked, garage and basement doors can be a significant vulnerability if not properly secured. Many garages have **thin, hollow doors** or **automatic openers** that are easy to disable. Similarly, basement doors might have old locks that haven't been upgraded in years.

- **Garage Door Security**: If your garage has an automatic opener, ensure that it has a **manual lock** that can be engaged in case of a power outage or malfunction. Use a **padlock** on the inside of the door to secure it from being lifted manually.
- **Reinforce Basement Doors**: Just like your exterior doors, basement doors should be solid-core and equipped with deadbolts. If the basement has windows, apply the same security measures as you would for ground-floor windows.

2.3 - Interior Defense: Creating Safe Rooms and Emergency Plans

While fortifying the exterior of your home is essential, you also need to plan for what happens **if an intruder gets inside**. Navy SEALs are taught to always have an escape route and to be ready to **fall back** to a more secure position if needed. In a bug-in scenario, this means creating **safe rooms** and having an emergency plan for when the outer defenses are breached.

Safe Rooms: Your Last Line of Defense

A **safe room** is a fortified area within your home where you can retreat in case of a break-in or attack. It's not just a panic room with steel doors and cameras (although that's an option); it can be any room that's been reinforced and stocked with supplies to keep you safe until help arrives or the threat is gone.

- **Choosing the Room**: The best safe rooms are located in the **center of the house**, away from windows and exterior walls. A bedroom, bathroom, or large closet can be converted into a safe room.
- **Reinforcing the Door**: Just like your exterior doors, the door to your safe room should be solid-core and equipped with a **deadbolt**. Consider adding a **security bar** or **brace** for extra protection.
- **Communication Devices**: Keep a **cell phone, radio**, or **walkie-talkie** in your safe room to maintain contact with the outside world. Ensure that you have a way to call for help, even if the power is out.
- **Stocking the Room**: Your safe room should be stocked with **food, water, medical supplies**, and **self-defense tools** (more on that below). This ensures that you can stay in the room for an extended period if needed.

Defense Tools: Preparing for Close-Quarters Combat

While it's always best to avoid direct confrontation, there may come a time during a bug-in when you need to **defend your home**. Navy SEALs are trained in

close-quarters combat (CQC), and many of the principles they use can be applied to home defense:

- **Firearms**: If you are comfortable with firearms, they can be an effective tool for home defense. A **shotgun** or **handgun** is generally best for close-quarters situations. Ensure that firearms are **secured** when not in use but readily accessible if needed.
- **Non-Lethal Weapons**: If you prefer non-lethal options, **pepper spray**, **tasers**, and **batons** can be effective at stopping an intruder without causing permanent harm.
- **Training**: No matter what weapons you choose, **training is essential**. Without proper training, a weapon can be more dangerous to you than to the intruder. Consider taking self-defense or firearms courses to learn how to use these tools effectively.

2.4 - Evacuation Planning: Knowing When to Bug Out

While the primary focus of this guide is on bugging in, there may come a time when your home is no longer safe, and you must **evacuate**. Navy SEALs always have a contingency plan, and so should you.

Creating an Evacuation Plan

Your evacuation plan should include:

- **Pre-determined Routes**: Identify multiple escape routes from your home and neighborhood. Roads may be blocked or unsafe, so consider **alternative routes** like back roads or footpaths.
- **Go Bags**: Each family member should have a **go bag** packed with essential supplies, including **food, water, clothing, medical supplies**, and **important documents**. The go bag should be stored in an easily accessible location, like near the front door or in the garage.
- **Rendezvous Points**: Establish **rendezvous points** where your family can meet if you are separated during the evacuation. Choose both a **nearby location** (like a neighbor's house) and a **distant location** (like a relative's house in another town).

2.5 - Psychological Barriers: Deterring Intruders Before They Act

Beyond the physical barriers, there are **psychological tactics** you can use to deter potential intruders. SEALs are trained to project **confidence and control**, and these same principles can be applied to home defense.

Signs and Warnings

Displaying **security signs** (even if you don't have a system) can make intruders think twice before attempting a break-in. Signs like "**Warning: Security**

Cameras in Use" or "**This Home Protected by Armed Response**" create the impression that your home is well-defended, even if it isn't fully equipped.

Dogs as Deterrents

Even if you don't own a dog, posting a sign that says "**Beware of Dog**" can be enough to scare off some intruders. If you do have a dog, their barking can alert you to potential threats and serve as a psychological deterrent for anyone considering entering your property.

Conclusion: Preparing for the Worst, Securing the Best

Securing your home during a bug-in scenario requires **layers of defense**, from the perimeter to the interior. By reinforcing entry points, creating safe rooms, planning for emergencies, and thinking like a Navy SEAL, you can transform your home into a fortress capable of withstanding any crisis. Remember, the key to survival is not just the tools you have, but the **mindset** and **preparation** you bring to the table.

Chapter 3: Mental Toughness: SEAL Psychology for Staying Calm in a Crisis

When you think of Navy SEALs, you likely envision their physical strength and endurance, the ability to push their bodies to the extreme in high-stress environments. However, what truly sets them apart is their **mental toughness**—the ability to stay focused, calm, and resilient under pressure. This chapter is dedicated to understanding the mindset required to survive a crisis while bugging in, applying SEAL principles to everyday life, and building the psychological fortitude necessary to endure long periods of isolation and uncertainty.

Mental toughness is about more than just gritting your teeth through a difficult situation. It's about maintaining **clarity, composure, and confidence** in the face of adversity. Whether you're dealing with natural disasters, societal collapse, or extended periods without electricity or food supplies, the way you handle stress can be the difference between success and failure. In a bug-in scenario, you won't just need to survive physically—you'll need to keep your mind sharp and your emotions in check, especially when the pressure is at its highest.

Navy SEALs train rigorously for mental endurance, honing their abilities to remain calm under fire, manage stress, and make quick decisions when lives are on the line. The good news is that the skills they use can be applied to anyone, in any situation. This chapter will walk you through the psychology of **mental toughness**, **coping mechanisms**, and **mindset shifts** that you can adopt to ensure you remain steady, resilient, and capable in a prolonged bug-in scenario.

3.1 - The SEAL Mindset: How Mental Toughness is Built

Mental toughness isn't something that SEALs are born with. It's something they build through years of training, exposure to stress, and learning to thrive in chaotic environments. The same can be true for you. You may not be going through Hell Week or getting dropped in a hostile environment with limited resources, but you can still apply the **same principles** to help you survive any crisis.

Commitment to Purpose

One of the core components of SEAL mental toughness is having a clear, **unshakable sense of purpose**. In military operations, SEALs are given a mission, and they are trained to focus relentlessly on completing that mission, regardless of the challenges they face. When bugging in, your mission is clear: protect your family, secure your home, and survive the crisis. By committing to this purpose, you eliminate any room for doubt or hesitation, focusing all your energy on what needs to be done.

- **Identify Your Mission**: Clearly define what your goals are for this bug-in situation. Whether it's keeping your family safe, maintaining food and water supplies, or staying informed about the crisis outside, knowing your mission will help you maintain focus.
- **Eliminate Distractions**: In a crisis, distractions can be dangerous. SEALs are trained to block out unnecessary noise and stay laser-focused on the task at hand. Apply this by avoiding distractions that might drain your mental energy, such as excessive social media or non-essential tasks.
- **Adaptability**: Mental toughness is also about flexibility. No plan is perfect, and in a long bug-in scenario, things may not go as expected. SEALs know that staying rigid in the face of changing circumstances can lead to failure. You need to adopt an **adaptive mindset**, ready to shift tactics and plans as new information becomes available or conditions worsen.

The Power of Grit and Perseverance

One of the defining characteristics of a Navy SEAL is their **unwillingness to quit**. This is often referred to as "grit"—the ability to persist through difficult conditions, push through physical and mental discomfort, and keep going when others would give up. When bugging in, you'll face long stretches of time where progress seems slow or conditions become harder. Grit will get you through those moments.

- **Small Victories**: SEALs are trained to break down large, overwhelming challenges into smaller tasks, focusing on achieving **small victories** along the way. When bugging in, celebrate small achievements: organizing supplies, reinforcing an entry point, or even just completing a daily routine. These victories provide mental momentum and a sense of accomplishment.
- **The 40% Rule**: SEALs are taught that when their body or mind feels like it's reached its limit, they're likely only operating at **40% of their true capacity**. This rule is a powerful tool for pushing through moments of fatigue or frustration during a crisis. When you feel like giving up, remind yourself that you likely have much more to give, and use this awareness to keep going.
- **Delaying Gratification**: Mental toughness often requires putting off short-term comfort for long-term gain. SEALs endure extreme

conditions because they know that the discomfort is temporary, and the mission's success is the ultimate reward. When bugging in, you might need to make sacrifices in the short term—whether it's rationing food, conserving energy, or postponing personal comforts. Understanding the bigger picture will help you endure these sacrifices without losing focus.

Stress Inoculation: Training the Brain to Handle Pressure

SEALs use a process known as **stress inoculation training (SIT)** to desensitize their minds and bodies to stress. This involves placing themselves in increasingly difficult and stressful situations so that when they encounter extreme stress in the real world, their body doesn't overreact. While you may not go through SEAL-style training, there are ways to condition yourself to handle stress more effectively during a bug-in.

- **Simulating Crisis Conditions**: One way to train your mind to handle stress is by simulating crisis conditions. Conduct **bug-in drills** with your family, turning off the electricity for a day or two, rationing food, and seeing how you handle the stress. By practicing under controlled circumstances, you prepare yourself for the real thing.
- **Breathing Techniques**: SEALs use **tactical breathing** to regulate their physiological

response to stress. By controlling your breathing, you can lower your heart rate, decrease adrenaline, and stay calm. Practice the **4x4 breathing technique**: inhale for four seconds, hold for four seconds, exhale for four seconds, and pause for four seconds. Repeat this cycle when you feel stressed.
- **Cognitive Reframing**: SEALs are trained to reframe how they view stress. Instead of seeing it as something negative, they learn to see it as a challenge that sharpens their performance. When bugging in, try to reframe stressful situations by focusing on what you can control and how the experience is making you stronger.

3.2 - Building Resilience: The SEAL Approach to Handling Setbacks

Resilience is the ability to recover from setbacks, keep moving forward, and maintain a positive outlook even in the face of adversity. SEALs are trained to endure failure, loss, and hardship without letting it break their spirit. In a prolonged bug-in, where supplies might run low, communication may fail, and uncertainty reigns, resilience becomes your most valuable mental asset.

Focus on What You Can Control

In a bug-in scenario, many aspects of the crisis will be completely beyond your control. Navy SEALs are trained to **focus on what they can control** and let go of what they can't. This mindset helps conserve mental energy and keeps you from wasting time on unproductive worry.

- **Actionable Steps**: When things start to feel overwhelming, break down the situation into **actionable steps**. Focus on what can be done in the present moment—whether it's securing your perimeter, rationing supplies, or checking on your family's well-being. Taking action helps restore a sense of control.
- **Letting Go of the Uncontrollable**: There will be many things you can't change: the duration of the crisis, external threats, or even the loss of certain comforts. SEALs learn to quickly identify what is outside of their control and let go of it mentally. Doing so helps reduce anxiety and keeps the mind clear for productive problem-solving.

The Power of Positivity and Visualization

While SEALs are some of the toughest fighters in the world, they're also trained to use **positive thinking** and **visualization techniques** to keep their spirits up and focus on success. In a long-term bug-in scenario, the mental drain can take a toll on your morale. Visualizing positive outcomes and maintaining a hopeful attitude can counterbalance the hardships you face.

- **Visualizing Success**: SEALs often use visualization to mentally rehearse success before a mission. When bugging in, take time each day to **visualize a successful outcome**—whether it's making it through a week without external supplies, or simply staying safe through the night. This helps the brain stay motivated and focused on achieving positive results.
- **Positive Reinforcement**: Even in the toughest situations, it's important to maintain some degree of **optimism**. While it might seem unrealistic to be overly positive in a crisis, acknowledging small wins and reminding yourself that things will eventually improve can help build psychological resilience. SEALs often use **self-talk** to reinforce this mindset, repeating positive affirmations like "I can do this" or "I will succeed" to stay mentally strong.

Embracing Hardship: Turning Discomfort Into Strength

In Navy SEAL training, recruits are subjected to extreme discomfort, deprivation, and pain. This is by design—it teaches them to embrace hardship and turn it into strength. The idea is that if you can endure the worst conditions, you'll be ready for anything. When bugging in, you'll face moments of discomfort—whether it's hunger, boredom, isolation, or fear. By learning to **accept and embrace discomfort**, you'll be better equipped to handle the emotional and physical strain of a long-term crisis.

- **The Cold Water Principle**: SEALs are often trained to withstand cold water, a metaphor for enduring difficult conditions. In a bug-in, you can apply this principle by embracing the challenges you face rather than resisting them. Whether it's rationing food, going without power, or dealing with isolation, accept the discomfort and remind yourself that each moment of hardship makes you mentally stronger.
- **Mental Conditioning**: Just as you condition your body through physical exercise, you can condition your mind by intentionally placing yourself in uncomfortable situations. Take on small challenges that push your limits—whether it's fasting for a day, limiting your use of technology, or spending time in silence. Each of these experiences builds mental resilience.

3.3 - Maintaining Morale: Mental Health Strategies for Long-Term Isolation

One of the biggest challenges during a bug-in scenario is dealing with **isolation and monotony**. SEALs often operate in isolation for long periods of time, separated from their teams or embedded in hostile environments. They are trained to maintain **morale**, **mental health**, and **focus**, even when the mission seems never-ending. In a bug-in, where you may be cut off from the outside world for extended periods, keeping your spirits up is just as important as securing supplies or reinforcing your home.

Creating Structure and Routine

One of the first things to break down in a crisis is routine. SEALs thrive on **structure**, even in chaotic environments. During a bug-in, creating and maintaining a daily routine helps preserve mental clarity and keeps the mind from becoming overwhelmed by the unknown.

- **Daily Tasks**: Set up a schedule for daily tasks such as meal preparation, home security checks, exercise, and communication. Structure gives you a sense of normalcy and control, even when the outside world is in chaos.
- **Family Roles**: If you're bugging in with family members, assign specific roles and responsibilities. SEAL teams function because every member has a role, and each role contributes to the success of the mission. Whether it's managing supplies, cooking, or monitoring the perimeter, giving each person a job helps maintain focus and purpose.
- **Physical Exercise**: Staying physically active is not only important for your health but also for your mental well-being. SEALs maintain rigorous physical fitness even in the most extreme conditions. In a bug-in, create a daily exercise routine that keeps you moving—whether it's

bodyweight exercises, stretching, or cardio. Exercise releases **endorphins**, which boost mood and reduce stress.

Managing Loneliness and Cabin Fever

Prolonged isolation can lead to feelings of **loneliness, cabin fever**, and even depression. SEALs are trained to cope with isolation by staying mentally engaged, maintaining communication, and focusing on the mission at hand. When bugging in, you may face similar psychological challenges, especially if you're cut off from friends, family, or external support.

- **Staying Connected**: If possible, maintain regular communication with loved ones outside of your home. Whether through **phone calls, radio**, or even writing letters, staying connected helps reduce feelings of isolation.
- **Mental Stimulation**: Keep your mind active by engaging in **mental exercises** or learning new skills. SEALs often use downtime to study, learn, or practice skills. In a bug-in, take up a **new hobby**, learn **survival skills**, or engage in **mental puzzles** to keep your brain sharp.
- **Creative Outlets**: Having a **creative outlet** can be a powerful way to manage the stress of isolation. Whether it's writing, drawing, or even organizing supplies in a new way, creativity helps you process emotions and stay mentally engaged.

Team Dynamics: Managing Stress in Group Settings

If you're bugging in with others, group dynamics can become a source of stress. SEALs operate in tightly knit teams, and they are trained to manage interpersonal relationships and group stress in high-stakes environments. In a bug-in scenario, especially over long periods, tension can rise between family members or housemates.

- **Conflict Resolution**: SEALs are trained to **resolve conflicts** quickly and efficiently to maintain team cohesion. In a bug-in, establish clear communication guidelines and address any conflicts as soon as they arise. Letting small grievances fester can lead to bigger problems down the line.
- **Group Morale**: Keeping group morale high is essential in a bug-in. Encourage **open communication**, celebrate small victories together, and ensure everyone has a role in maintaining the home. Group activities, like shared meals or watching a movie (if power is available), help maintain unity.

Conclusion: The SEAL Blueprint for Mental Toughness

Mental toughness is not about being emotionless or invulnerable. It's about building the **mental resilience** to face challenges head-on, adapt to changing circumstances, and maintain your mission focus even when the world seems to be falling apart. By adopting the **SEAL mindset**—focusing on what you can control, embracing discomfort, and keeping morale high—you'll be well-prepared to handle the psychological challenges of a long-term bug-in.

This chapter lays the groundwork for developing **psychological fortitude**, helping you endure the trials of isolation, scarcity, and stress that may accompany a crisis. Remember, survival isn't just about the physical—it's about having the mental tools to thrive under pressure.

Chapter 4: Surviving Long-Term Power Outages: Adapt, Overcome, and Thrive

A long-term power outage is one of the most common and disruptive challenges you may face during a crisis. Whether it's caused by natural disasters, widespread civil unrest, or a breakdown of infrastructure, the loss of electricity can turn your world upside down. Everything from heating and cooling to communication, cooking, and water supply depends on power. In this chapter, we'll dive into the critical steps necessary to survive and thrive during a prolonged power outage, drawing on **Navy SEALs' adaptability, resourcefulness, and focus** on preparation.

Unlike a short-term blackout, a long-term power outage can last weeks or even months, leaving you without the modern conveniences you're accustomed to. Navy SEALs are trained to operate in the most austere environments, where they often have no access to electricity or other basic necessities. The tactics they use to **adapt, overcome**, and **thrive** in such conditions are applicable to anyone preparing to bug in during a power outage. This chapter will guide you through how to prepare your home, find alternative power sources, stay warm or cool, and ensure you have access to essential resources when the grid is down.

4.1 - Understanding the Impact of a Long-Term Power Outage

The first step in surviving a long-term power outage is understanding just how significantly it will affect your daily life. The moment the power goes out, you lose more than just light and heat—you lose access to many vital systems that are part of your modern lifestyle. In a bug-in scenario, your ability to survive the loss of these systems will depend on how well-prepared you are.

Common Immediate Effects

- **Loss of lighting**: In the absence of natural light, your home will be in darkness. This limits visibility and can create safety hazards.
- **Temperature control**: Without power, heating and cooling systems will fail, which can be especially dangerous during extreme weather conditions.
- **Food spoilage**: Without refrigeration, perishable food items will begin to spoil within hours.
- **Water supply issues**: If your home relies on an electric pump for water, you'll lose access to your water supply.
- **Communication disruption**: Internet and phone service may go down if cell towers and data

centers lose power, leaving you isolated from the outside world.
- **Security risks**: Home security systems, including alarms and cameras, often rely on electricity, leaving your home vulnerable.

In Navy SEAL operations, failure to anticipate and prepare for these types of disruptions can be the difference between life and death. Similarly, your ability to plan ahead for the effects of a power outage is critical to your survival during a bug-in.

Developing a Power-Outage Survival Plan

Your **Power-Outage Survival Plan** should take into account the specific vulnerabilities of your home and family. To begin, you need to assess your reliance on electricity and prioritize which systems or services are most critical for survival. This plan should cover the following key areas:

- **Lighting**: Safe, alternative lighting sources for both indoor and outdoor use.
- **Temperature control**: Methods for staying warm in the winter and cool in the summer without access to HVAC systems.
- **Water and food**: Ensuring a stable water supply and the preservation of food.
- **Communication and information**: Keeping in contact with the outside world and staying informed about the status of the crisis.

- **Security**: Maintaining a secure perimeter even if security systems fail.

By having a detailed plan in place, you'll ensure that when the power goes out, you're not caught off guard and can transition smoothly into survival mode.

4.2 - Lighting Solutions for a Grid-Down Scenario

In a power outage, one of the first things you'll lose is lighting. Darkness can cause accidents, hamper your ability to perform basic tasks, and increase feelings of fear and insecurity. SEALs are trained to operate in low-visibility environments and use both **natural and artificial light** to their advantage. Here's how you can prepare your home for long-term darkness:

Battery-Powered Lanterns and Flashlights

While it might seem obvious, having an adequate supply of **battery-powered lanterns** and **flashlights** is critical. The key is ensuring you have enough light sources for each member of your household and that they are **easily accessible** in the event of an outage.

- **Lanterns**: Lanterns provide ambient lighting, making them ideal for lighting up a room or outdoor area. Choose models with **long battery life** and consider those with adjustable brightness settings to conserve power.

- **Flashlights**: Every family member should have their own flashlight. Choose tactical-grade flashlights that are **durable**, **waterproof**, and have a high **lumen output** for maximum visibility in dark areas.

Solar-Powered and Crank Lanterns

Because batteries may become scarce during a prolonged outage, invest in **solar-powered** and **hand-crank lanterns**. These lanterns don't rely on batteries, making them ideal for long-term use:

- **Solar Lanterns**: These can charge during the day and provide hours of light at night. Place them outside during daylight hours to absorb energy and bring them inside when the sun goes down.
- **Crank Lanterns**: These can be powered manually by cranking a handle. While not as convenient as solar lanterns, they're an excellent backup when sunlight isn't available.

Candles and Oil Lamps

While modern light sources are preferable, **candles** and **oil lamps** can be useful in a prolonged outage. SEALs are trained to minimize risk in all situations, so it's important to use these **open flame light sources** with caution:

- **Candles**: Use candles as a **last resort** when other lighting options are unavailable. Be mindful of the fire risk, and never leave candles unattended.
- **Oil Lamps**: Oil lamps can provide more stable light than candles and burn for longer periods of time. Make sure you have a **supply of lamp oil** stored safely, as well as **matches or lighters** for ignition.

Glow Sticks and Chemical Lights

In a pinch, **glow sticks** or **chemical lights** can provide several hours of light without the need for batteries or electricity. SEALs use chemical lights in tactical environments because they're lightweight, reliable, and produce **no noise or heat**, making them a safe option for illuminating small areas without attracting unwanted attention.

4.3 - Temperature Control: Staying Warm or Cool Without Power

One of the most critical challenges of a long-term power outage is **maintaining a livable temperature** inside your home. SEALs are trained to endure harsh climates, from the freezing cold of winter warfare training in Kodiak, Alaska, to the sweltering heat of desert operations. When bugging in during a power outage, you must be prepared to **adapt to the weather** using low-tech and no-tech solutions.

Staying Warm in Cold Climates

Cold weather can be deadly during a power outage, especially if your home's heating system is electrically powered. Without preparation, hypothermia becomes a very real risk. Here's how to keep your home warm and stay comfortable:

- **Layering Clothing**: SEALs use layering to regulate body temperature during extreme cold. You should wear multiple layers of loose-fitting clothing to trap warm air. Start with a **moisture-wicking base layer** to keep sweat off your skin, followed by insulating layers like **fleece**, and finally, a wind-resistant outer layer if necessary.
- **Insulating Your Home**: Prevent heat loss by **sealing off unused rooms** and concentrating heat in a smaller area. Use heavy blankets or **tape up plastic sheeting** to block drafts from doors and windows. In SEAL survival training, insulating your shelter is one of the first steps in cold-weather survival, and your home should be treated the same way.
- **Sleeping Bags**: High-quality **cold-weather sleeping bags** are a must-have for prolonged power outages. A sleeping bag rated for **sub-zero temperatures** can keep you warm even in the coldest environments. If you're unable to heat your home, **sleeping in your bag**

will retain much more warmth than using regular blankets.
- **Emergency Blankets**: SEALs carry **mylar emergency blankets** in their kits because they are lightweight, compact, and highly effective at reflecting body heat. Stockpile emergency blankets and use them in combination with other blankets to trap body heat.

Portable Heaters and Fireplaces

If you have a **fireplace** or **wood stove**, you have a major advantage during a power outage. Fireplaces can be used not only for heating but also for cooking and boiling water. However, make sure to have the following in place:

- **A Stockpile of Wood**: Ensure you have enough **dry firewood** to last for the duration of the outage. Wood should be stored in a dry, accessible location. If you rely on a wood stove, make sure you have **fire-starter materials** such as kindling, matches, and lighter fluid.
- **Ventilation**: Ensure proper **ventilation** when using a wood stove or fireplace to avoid carbon monoxide buildup. Even in survival situations, SEALs prioritize safety, and you should too.
- **Propane or Kerosene Heaters**: If you don't have a fireplace, **propane** or **kerosene heaters** can provide temporary heat. However, they must be used with **extreme caution**. Always ensure

proper ventilation, and never leave heaters running unattended.

Staying Cool in Hot Climates

On the other end of the spectrum, staying cool in **hot weather** without air conditioning can also pose serious health risks. **Heat exhaustion** and **dehydration** can be fatal if you're not prepared to deal with rising temperatures. SEALs learn to keep their core temperature under control in desert environments, and you can apply the same principles to your home.

- **Shade and Ventilation**: Keep your home cool by **blocking direct sunlight** with curtains, blinds, or reflective materials. SEALs use **natural shade** to avoid overheating, and you should do the same by closing off sun-exposed areas. Open windows on opposite sides of the house to create **cross-ventilation** and allow air to flow.
- **Hydration**: Staying hydrated is critical in hot climates. SEALs are trained to drink water regularly to avoid heat exhaustion, and during a power outage, you should do the same. Keep an eye on your **urine color**—it should be pale or clear, a key indicator of proper hydration.
- **Cooling Stations**: Create **cooling stations** in shaded or cooler parts of your home. Use **damp towels** on your skin or take cool showers to bring down your core temperature. In the absence of power, even something as simple as

misting your skin with water can be an effective cooling method.

4.4 - Ensuring a Steady Water Supply

Water is essential for survival, and during a power outage, your access to clean water may be disrupted, especially if your water supply relies on **electric pumps** or if local treatment plants are down. SEALs are trained to find, purify, and store water in the field, and you'll need to take similar steps to ensure your family has enough drinking water during a long-term outage.

Storing Water for Emergencies

The first and most important step is to store a **sufficient supply of water** before the power goes out. The rule of thumb is to store **one gallon per person per day** for drinking, cooking, and hygiene. Aim for at least a **two-week supply** for each member of your household, but ideally, you should store enough to last a month or more. Use the following storage methods:

- **Large Water Containers**: Use **55-gallon drums** or **five-gallon water jugs** for bulk water storage. These containers are food-grade and designed for long-term storage.
- **Water Bottles**: Store commercially bottled water as a backup. These bottles are sealed and can last for years if kept in a cool, dark place.

Alternative Water Sources

If your stored water runs out, you'll need to find **alternative water sources** and purify them for safe drinking. Navy SEALs often rely on natural water sources during missions, and you can use similar tactics at home:

- **Rainwater Collection**: Use **rain barrels** to collect water from your roof or other surfaces. Make sure your collection system is properly filtered to remove debris and contaminants before drinking.
- **Lakes, Rivers, and Streams**: If you live near a natural water source, you can collect water from lakes, rivers, or streams. However, this water must be purified before drinking to eliminate bacteria, viruses, and parasites.

Purifying Water

Water purification is critical in a long-term outage, as untreated water can carry harmful pathogens. SEALs are trained to purify water in the field, and you can apply the same methods at home:

- **Boiling Water**: Boiling is one of the simplest and most effective ways to purify water. Bring water to a **rolling boil** for at least **one minute** to kill bacteria, viruses, and parasites.
- **Water Filters**: Use portable **water filters** like the **LifeStraw** or **Sawyer Mini**, which can remove

most pathogens. These filters are compact and ideal for both individual use and larger family needs.
- **Water Purification Tablets**: Keep a stock of **water purification tablets**, which are easy to store and use in an emergency. These tablets typically contain **chlorine dioxide** or **iodine**, both of which are effective at killing harmful microorganisms.

4.5 - Food Preservation and Cooking Without Power

One of the biggest challenges during a long-term power outage is keeping your food from spoiling and finding ways to cook it. Navy SEALs learn to survive on minimal rations and often rely on preserved or lightweight foods in the field. You'll need to take a similar approach to food storage and meal preparation in your bug-in scenario.

Preserving Perishable Foods

When the power goes out, your refrigerator and freezer will lose the ability to keep food at safe temperatures. Here's how to preserve your food as long as possible:

- **Freeze Water Jugs**: If you know a power outage is likely, **freeze several water jugs** ahead of time. Place these jugs in your refrigerator and freezer to act as **ice packs**, keeping food colder for longer.

- **Dry Ice or Ice Blocks**: In longer outages, **dry ice** or **block ice** can be used to keep your freezer cold. Just be cautious when handling dry ice, as it can cause burns.
- **Preserved Foods**: Focus on **non-perishable foods** like canned goods, freeze-dried meals, and dried fruits or meats. SEALs carry **MREs (Meals Ready-to-Eat)** in the field because they're lightweight, have long shelf lives, and don't require refrigeration.

Alternative Cooking Methods

Without electricity, you'll need to rely on **alternative cooking methods** to prepare your meals. SEALs are trained to use whatever tools are available to them, and you can do the same with these simple techniques:

- **Camp Stoves**: A **propane** or **butane camp stove** is one of the best tools for cooking during a power outage. These stoves are portable, easy to use, and allow you to cook indoors (with proper ventilation) or outdoors.
- **Fire Pits**: If you have a fire pit in your yard, you can use it to cook over an open flame. Just make sure to have a **cast-iron skillet** or **grill grate** that can withstand high temperatures.
- **Solar Ovens**: A **solar oven** is a fantastic tool for long-term outages, as it uses the sun's energy to cook food. Solar ovens work best in direct

sunlight and can reach temperatures high enough to bake or boil.

4.6 - Staying Informed: Communication and Information in a Power Outage

One of the most disorienting aspects of a long-term power outage is the lack of information. Without the internet, TV, or radio, you might feel cut off from the outside world. Navy SEALs are trained to maintain communication and stay informed, even in the most isolated environments. Here's how you can do the same:

Emergency Radios

An **emergency radio** is an essential tool for staying informed during a power outage. Choose a radio that can pick up **AM/FM signals** as well as **NOAA weather channels**. Look for radios that are **battery-powered, solar-powered, or hand-crank** to ensure you can still receive updates even without electricity.

Two-Way Radios

If you're bugging in with a group, two-way radios (also known as **walkie-talkies**) can be invaluable for keeping in touch if cell service goes down. Make sure to have extra batteries on hand, and practice using them with your family so everyone knows how to communicate efficiently.

Staying Mentally Prepared

Being cut off from the flow of information can cause anxiety and fear. SEALs learn to manage these emotions by focusing on **what they can control** and staying mentally resilient. Focus on keeping yourself and your family safe, and trust that the crisis will pass.

Conclusion: Adapting to Life Without Power

A long-term power outage can be one of the most challenging aspects of any bug-in scenario. However, with the right preparation, resourcefulness, and mindset, you can survive—and even thrive—without electricity. By following the principles of **adaptability** and **resilience** that SEALs use in the field, you'll be well-equipped to handle whatever challenges arise during an extended blackout.

The next chapter will dive into another critical resource: **water**—how to secure, purify, and store it to ensure your survival.

Chapter 5: Water Supply and Purification: Navy SEAL Hydration Tactics

Water is the foundation of survival. The human body can survive for weeks without food, but only a few days without water. In a long-term bug-in scenario, ensuring a safe and sustainable water supply becomes critical. This chapter is dedicated to showing you how to secure, store, and purify water, using strategies based on **Navy SEAL survival tactics**. SEALs operate in some of the most challenging environments in the world, often without access to clean water, and must rely on both natural sources and purification techniques to stay hydrated.

Whether you're facing a power outage, civil unrest, or natural disaster, your access to clean drinking water may be compromised. Municipal water systems can fail, wells can go dry, and bottled water can run out. Being prepared with a long-term water plan is crucial for your bug-in strategy. You'll need to know how to secure water, how much to store, and how to purify it to ensure it's safe for consumption.

5.1 - Assessing Your Water Needs

Before diving into the specifics of water storage and purification, it's essential to understand **how much**

water you actually need during a bug-in scenario. SEALs are trained to calculate their water needs based on mission length, physical activity, and environmental conditions, and the same principles apply to your survival at home.

How Much Water Do You Need?

The general rule of thumb for water consumption in a survival scenario is **one gallon per person per day**. This gallon includes water for drinking, cooking, and basic hygiene. However, your specific water needs will vary depending on a few factors:

- **Climate**: In hotter climates, you'll need more water to stay hydrated and regulate your body temperature. Physical exertion in high heat can double your daily water requirement.
- **Physical Activity**: If you're engaging in physically demanding activities, such as reinforcing your home or performing maintenance, your body will require more water.
- **Health Conditions**: Certain health conditions, such as diabetes or kidney issues, may increase your water needs. Consider the unique health requirements of each member of your household.

For planning purposes, you should store at least a **two-week supply of water** per person. This equates to **14 gallons per person** for basic survival, but for long-term crises, it's wise to plan for a month or more of water reserves.

5.2 - Storing Water for Long-Term Survival

Navy SEALs often operate in areas where they must carry their own water supplies for days or weeks at a time. This requires careful planning and efficient storage solutions. In a bug-in scenario, you have the advantage of being able to store larger quantities of water, but it must be done safely and efficiently.

Choosing Water Storage Containers

The containers you use to store water are just as important as the water itself. Poorly stored water can become contaminated, leading to illness or even death. SEALs rely on **durable, safe water storage systems** that protect their water from environmental hazards, and you should do the same at home.

- **Food-Grade Plastic Containers**: Look for **food-grade plastic containers** designed for long-term water storage. These containers are BPA-free and won't leach harmful chemicals into your water. Popular options include **55-gallon drums**, which can store large amounts of water, and **five-gallon jugs**, which are easier to transport.

- **Water Bricks**: These stackable **water bricks** are another excellent option for long-term storage. They are durable, space-efficient, and can be easily moved when necessary.
- **Collapsible Containers**: In situations where space is limited, **collapsible water containers** are ideal. They can be stored flat when not in use and expanded as needed. These are also great for collecting rainwater in emergencies.

Sanitizing Water Storage Containers

Before storing water, it's crucial to **sanitize** your containers to prevent contamination. Even food-grade containers can harbor bacteria or chemical residues if not properly cleaned. Here's how to sanitize your water storage containers:

1. **Wash the containers** with hot, soapy water. Rinse thoroughly.
2. Prepare a **bleach solution** using **one teaspoon of unscented liquid household bleach per quart of water**.
3. Swirl the bleach solution inside the container, making sure it contacts all surfaces.
4. Rinse with clean water and allow the container to dry completely before filling it with drinking water.

Filling and Storing Water

When filling your containers, always use **clean, potable water** from a trusted source. If you're filling from tap water, you can add a small amount of unscented bleach (two drops per quart) to prevent bacterial growth during long-term storage. Once filled, store your water in a **cool, dark place**, away from direct sunlight and extreme temperatures. Sunlight and heat can degrade plastic containers and promote the growth of algae or bacteria.

- **Rotate Stored Water**: Even properly stored water should be rotated every **six months to one year** to ensure freshness. Label each container with the date it was filled to keep track.

Water Storage for Non-Drinkable Uses

In addition to storing water for drinking and cooking, it's wise to store **non-potable water** for tasks like cleaning, bathing, and flushing toilets. You can store non-potable water in **rain barrels** or other large containers, but it should be clearly marked as non-drinkable to avoid confusion.

5.3 - Alternative Water Sources During a Crisis

Navy SEALs are trained to locate and extract water from a variety of natural sources, even in the most inhospitable environments. During a long-term bug-in, your stored water may eventually run out, and you'll need to tap into **alternative water sources**. Knowing how to find and purify water from **natural and man-made sources** is a critical survival skill.

Rainwater Collection

Rainwater is one of the most accessible alternative water sources, and it's free. SEALs are taught to collect rainwater using tarps or other materials, and you can apply the same techniques to your home.

- **Rain Barrels**: Install **rain barrels** at the base of your home's gutters to collect water runoff during storms. Make sure your gutters are clean and free of debris, and install **mesh covers** on the rain barrels to keep out insects and leaves.
- **Tarp Collection**: If you don't have a rain barrel system, you can still collect rainwater using a simple **tarp and container setup**. Stretch a tarp between two points, allowing rainwater to pool in the center and run into a clean container.

Rainwater collected from roofs or other surfaces should always be **purified before drinking**, as it can contain harmful chemicals, bird droppings, or other contaminants.

Nearby Natural Water Sources

If you live near a **lake, river, or stream**, these can be invaluable sources of water during a bug-in. However, these natural sources are often contaminated with bacteria, parasites, or industrial runoff, making purification essential.

- **Surface Water Collection**: When collecting water from a natural source, try to **gather from the cleanest part** of the water body, away from potential pollutants. In moving water, collect water from the **upstream side** to avoid contamination from downstream.

Well Water

If your home has a **well**, you may be able to continue using well water during a power outage, provided you have a way to power the pump. SEALs often rely on well water in rural locations, and it can be a reliable source during a crisis.

- **Manual Well Pumps**: Consider installing a **manual well pump** that doesn't require electricity. This allows you to draw water from the well even if the power is out.
- **Solar-Powered Pumps**: Alternatively, you can invest in a **solar-powered well pump**, which uses energy from the sun to provide a continuous supply of water during outages.

5.4 - Purifying Water for Safe Drinking

Water collected from rain, natural sources, or even some stored water can be unsafe to drink without proper purification. Navy SEALs use a variety of purification methods in the field, including **boiling, chemical treatment, and filtration**, all of which are highly effective for bug-in scenarios.

Boiling Water

Boiling is one of the simplest and most effective ways to purify water, killing **bacteria, viruses, and parasites**. It's a technique SEALs rely on when other purification methods aren't available, and it's something you can easily do at home.

- **How to Boil Water**: Bring water to a **rolling boil** for at least **one minute** (or three minutes at higher altitudes) to ensure all pathogens are killed.
- **Storage After Boiling**: Once boiled, let the water cool before pouring it into a clean, sanitized container for storage. Boiled water should be consumed within **24-48 hours** to ensure freshness.

Chemical Water Purification

Chemical treatment is another effective method for purifying water, especially in situations where boiling isn't practical. SEALs often carry **water purification tablets** as part of their field kits, and you should have a supply at home as well.

- **Water Purification Tablets**: These tablets typically contain **chlorine dioxide** or **iodine**, both of which are highly effective at killing harmful microorganisms. Follow the instructions on the packaging for proper dosage and wait times.
- **Bleach**: In the absence of tablets, **unscented household bleach** can be used to purify water. Add **two drops of bleach per quart of water**, stir, and let it sit for **30 minutes** before drinking. The water should have a faint chlorine smell, which indicates it's been properly treated.

Water Filtration

Water filters are another vital tool for ensuring your drinking water is safe. SEALs use portable filters in the field, and you can use similar filtration systems at home.

- **Portable Water Filters**: Systems like the **LifeStraw**, **Sawyer Mini**, or **Katadyn** filters are lightweight, portable, and effective at removing bacteria, protozoa, and other pathogens. These filters are ideal for purifying water on the go, but they can also be used in the home during a long-term bug-in.
- **Gravity Filters**: For larger-scale filtration, consider a **gravity-fed water filter** like the **Berkey** system. These filters are designed for

long-term use and can purify gallons of water per day, making them an excellent choice for family water needs during a crisis.

Solar Water Disinfection (SODIS)

In a pinch, you can purify water using the **SODIS method**, which relies on the sun's UV rays to kill harmful microorganisms. This method is often used in remote areas by SEALs and other survivalists when no other options are available.

- **How to Use SODIS**: Fill a **clear plastic or glass bottle** with water and place it in direct sunlight for at least **six hours**. The UV rays from the sun will kill most bacteria, viruses, and protozoa. This method works best in bright sunlight and should only be used with clear water, as murky or cloudy water reduces its effectiveness.

5.5 - Maintaining Hygiene and Sanitation with Limited Water

In a long-term bug-in scenario, your water supply will need to stretch beyond drinking and cooking. Hygiene and sanitation are crucial for preventing illness, especially in situations where medical help may be limited. SEALs maintain strict hygiene even in the harshest environments, and you should do the same.

Conserving Water for Hygiene

During a power outage or crisis, you'll need to conserve water while still maintaining basic hygiene standards. Here are some tips for managing personal hygiene with a limited water supply:

- **Sponge Baths**: Instead of taking full showers, use a **damp cloth or sponge** to wipe down your body, focusing on areas prone to sweat and bacteria, like underarms, feet, and groin.
- **Hand Washing**: Washing your hands is critical for preventing the spread of disease, especially after using the bathroom or handling food. Use **hand sanitizer** when water is scarce, or wash with a small amount of soap and water from a basin.
- **Brushing Teeth**: Use a small cup of water to brush your teeth, conserving as much water as possible. SEALs often use minimal amounts of water in the field, and you can do the same by rinsing your mouth with just a sip of water after brushing.

Sanitation and Waste Management

Maintaining proper sanitation is essential to prevent the spread of disease. In Navy SEAL operations, sanitation is prioritized to keep the team healthy and operational. Here's how to manage waste and maintain hygiene in a long-term bug-in:

- **Toilet Alternatives**: If your plumbing is out of commission due to a power outage, you can create a simple **bucket toilet**. Use a **5-gallon bucket lined with heavy-duty trash bags**. After each use, sprinkle a bit of **cat litter** or **sawdust** to control odors and absorb liquids.
- **Waste Disposal**: Proper disposal of human waste is critical for hygiene. Double-bag your waste and store it in a sealed container until it can be safely disposed of, or bury it at least 200 feet from your home and any water sources.
- **Cleaning Surfaces**: Use **disinfectant wipes** or a diluted bleach solution to regularly clean surfaces, particularly in food preparation areas and bathrooms. Keeping your home clean will reduce the risk of illness.

5.6 - Final Thoughts: Water is Life

As Navy SEALs are trained to know, **water is life**. In a long-term bug-in scenario, your survival hinges on your ability to secure, store, and purify water. By implementing the strategies and tactics outlined in this chapter, you'll be prepared to meet your family's water needs, no matter how long the crisis lasts. Remember, preparation is key—ensuring your water supply is secure before disaster strikes will give you a significant advantage when the unexpected happens.

Chapter 6: Food Storage for Extended Bug-Ins: What SEALs Eat to Survive

In a long-term bug-in scenario, having access to food is as critical as securing water. When grocery stores run out of stock or supply chains are disrupted, you will need to rely on stored food and creative meal preparation to sustain yourself and your family. Just like Navy SEALs, who are trained to survive on minimal rations in harsh environments, you will need to know how to store, preserve, and prepare food that can last for weeks, months, or even years.

This chapter will guide you through the fundamentals of **food storage**, the types of food that are ideal for long-term bug-ins, and how to ration and stretch your supplies. We'll explore how SEALs maintain their energy in survival situations and how you can adapt similar principles to your home preparedness plan. The key to long-term survival is not only stockpiling the right types of food but also knowing how to manage them effectively to avoid waste and ensure that you have enough to last.

6.1 - Assessing Your Food Needs: How Much Should You Store?

Before diving into the specifics of food storage, the first step is to calculate **how much food you need** for an extended bug-in scenario. Navy SEALs are taught to assess their food needs based on the **length of the mission**, their **physical activity level**, and **environmental factors**. The same calculations apply when planning your food storage for a bug-in.

Daily Caloric Requirements

The average adult needs **2,000 to 2,500 calories per day** to maintain normal bodily functions, but in a crisis, this number can vary based on several factors:

- **Physical Activity**: If you are performing physically demanding tasks like securing your home, chopping wood, or repairing infrastructure, your caloric needs will increase. Active adults may require up to **3,000 calories per day**.
- **Environmental Conditions**: Cold weather environments require more energy to maintain body heat, so your caloric intake will need to increase in colder climates.
- **Special Dietary Needs**: Take into account any **special dietary needs** for your household, such as for children, the elderly, or individuals with medical conditions that require specific types of food.

Calculating Total Food Storage Needs

To ensure you have enough food, plan to store a **three-month supply** of non-perishable food items. A simple formula for determining how much food you need is to multiply your daily caloric requirement by the number of people in your household and the number of days you plan to bug-in:

- **Example**: For a family of four with an average daily caloric requirement of 2,200 calories, you would need:
    - **2,200 calories x 4 people x 90 days = 792,000 total calories** stored.

Once you've calculated the total caloric needs for your household, you can start planning the types of food to store and how to distribute those calories across different categories.

6.2 - Types of Food to Store: Building a Balanced Stockpile

When planning for a bug-in, the goal is to create a stockpile that is both **calorically sufficient** and **nutritionally balanced**. SEALs rely on a mix of **high-calorie, nutrient-dense foods** to maintain energy during missions, and the same principle applies to long-term survival at home. You will need to store foods that provide a good balance of **carbohydrates, protein, fats**, and **vitamins**.

Non-Perishable Foods for Long-Term Storage

Here are the main categories of non-perishable foods to focus on when building your stockpile:

1. **Grains and Carbohydrates**
    - **Rice**: One of the best staples for long-term storage, rice is affordable, versatile, and has a long shelf life. Store both **white rice** and **brown rice** (although brown rice has a shorter shelf life due to its higher oil content).
    - **Pasta**: Dry pasta is another excellent carbohydrate source with a long shelf life. It's easy to cook and pairs well with various other preserved foods.
    - **Oats and Cereal Grains**: **Oats**, **barley**, and **quinoa** are nutritious grains that are easy to store and provide a substantial caloric intake.
    - **Flour**: **Wheat flour** and other types of flour (like cornmeal or almond flour) can be stored for baking and meal preparation. Flour has a shorter shelf life compared to other dry goods, so consider storing **whole grains** and grinding them as needed.
2. **Proteins**
    - **Canned Meat**: Stock up on canned proteins such as **tuna, chicken, beef**, and **salmon**. These are excellent sources

of protein that require no refrigeration and can be eaten as-is or used in recipes.
- **Dried Meat**: **Jerky** or **dehydrated meats** are great for long-term storage. They are lightweight, calorie-dense, and provide a good protein source.
- **Beans and Legumes**: **Dried beans** (like black beans, lentils, and chickpeas) are a fantastic source of both protein and fiber. They have an incredibly long shelf life and can be used in soups, stews, or on their own.
- **Canned Beans**: In addition to dried beans, store **canned beans** as a convenient alternative. They require no soaking and can be eaten cold if needed.

3. **Fats**
    - **Cooking Oils**: Store long-lasting oils like **olive oil**, **coconut oil**, and **ghee**. Fats are essential for energy and help improve the flavor of meals. Keep oils in a cool, dark place to extend their shelf life.
    - **Peanut Butter**: Peanut butter is an excellent source of protein and healthy fats. It has a long shelf life and is versatile

enough to eat with other food items like bread, crackers, or rice cakes.
- **Nuts and Seeds**: **Almonds, peanuts, sunflower seeds**, and **walnuts** are calorie-dense foods that provide fats, protein, and micronutrients.

4. **Fruits and Vegetables**
    - **Canned Vegetables**: While fresh produce won't last in a long-term bug-in, canned vegetables (like **corn, peas, carrots**, and **tomatoes**) are essential for maintaining balanced nutrition.
    - **Canned Fruits**: Store canned fruits (like **pineapple, peaches, apples**, and **pears**) to provide vitamins and a source of natural sugars.
    - **Dried Fruits**: **Raisins, cranberries, apricots,** and **dates** are packed with vitamins and minerals and have a long shelf life. They are also lightweight and can be added to meals for flavor and nutrition.

5. **Dairy Alternatives**
    - **Powdered Milk**: **Non-fat powdered milk** is an excellent source of calcium and can be used in cooking or for drinking. It has a long shelf life and can be reconstituted with water.
    - **Shelf-Stable Milk**: **UHT (ultra-high temperature)** processed milk is a good

alternative to fresh milk, as it can be stored at room temperature until opened.
6. **Miscellaneous Staples**
    - **Salt and Spices**: Don't forget to stock up on **salt, pepper, sugar**, and a variety of **spices**. These will be critical for making bland food taste better and providing flavor to otherwise basic meals.
    - **Honey**: Honey is a natural sweetener with an indefinite shelf life. It can be used in a variety of recipes or as a quick energy source.
    - **Vinegar**: Store **white vinegar** and **apple cider vinegar**, both of which have a long shelf life and can be used for cleaning, food preservation, and cooking.

Freeze-Dried and Dehydrated Foods

Freeze-dried foods are some of the best options for long-term storage because they maintain their nutritional content and taste while being lightweight and easy to store. SEALs often carry **freeze-dried meals** during extended operations due to their durability and convenience.

- **Freeze-Dried Meals**: Companies like **Mountain House** and **Wise Company** offer freeze-dried meals that can last up to **25 years** in storage. These meals only require hot water to prepare

and offer a balanced combination of calories, protein, and nutrients.
- **Dehydrated Fruits and Vegetables**: Dehydrating fruits and vegetables removes the water content, making them lightweight and easy to store for years. They can be rehydrated in soups or eaten as-is for snacks.

MREs (Meals Ready to Eat)

Navy SEALs often rely on **MREs** during missions where traditional cooking methods are impractical. MREs are complete meals that come in vacuum-sealed packages and are designed to last for years without refrigeration. While MREs can be expensive, they are an excellent addition to your emergency food supply because they are pre-cooked and can be eaten hot or cold.

6.3 - Food Preservation Methods: Extending Shelf Life

While some foods are naturally long-lasting, others will require additional preservation methods to ensure they remain safe to eat throughout a long-term crisis. SEALs are trained in food preservation techniques to make the most of limited resources, and you can use similar methods to extend the shelf life of your food supply.

Canning and Jarring

Home canning is an excellent way to preserve fruits, vegetables, and even meats for long-term storage. When done correctly, canned foods can last for **years** without refrigeration.

- **Water Bath Canning**: This method is suitable for high-acid foods like fruits, tomatoes, pickles, and jams. The process involves sealing food in jars and submerging them in boiling water for a specific period of time to kill bacteria and create a vacuum seal.
- **Pressure Canning**: Pressure canning is used for **low-acid foods** such as vegetables, meats, and soups. A pressure canner raises the temperature above boiling, ensuring that harmful bacteria like botulism are eliminated.

Dehydration

Dehydration is one of the oldest methods of food preservation, and it's highly effective for extending the shelf life of fruits, vegetables, and meats. Dehydrated foods take up less space than fresh or canned foods, making them ideal for bug-in scenarios.

- **Electric Dehydrators**: Using an electric dehydrator allows you to dry food evenly and quickly. Dehydrated fruits like apples, bananas, and mangoes make excellent snacks, while dried vegetables can be rehydrated in soups and stews.

- **Sun Drying**: If you don't have an electric dehydrator, you can also use the **sun drying method**. Lay thinly sliced fruits or vegetables on a drying rack, cover with a cloth to keep out insects, and place them in a sunny, well-ventilated area for several days.

Fermentation

Fermentation is a preservation method that can also enhance the nutritional value of foods. **Sauerkraut**, **kimchi**, **pickles**, and **yogurt** are all examples of fermented foods that can be stored for long periods of time without refrigeration.

- **How to Ferment**: Fermentation is a process in which natural bacteria convert sugars into acids or alcohol, which acts as a preservative. To make sauerkraut, for example, you can pack cabbage tightly into a jar with salt, and over time, the cabbage will ferment in its own juices. Fermented foods are rich in **probiotics**, which can help maintain digestive health during a crisis.

Salting and Smoking

Both **salting** and **smoking** are traditional preservation techniques used to cure meats and fish. Navy SEALs, operating in remote locations, often rely on these techniques to extend the life of fresh catches.

- **Salting**: Salt draws moisture out of food, making it inhospitable to bacteria. **Curing meats** with salt is a reliable way to preserve protein for months or even years.
- **Smoking**: Smoking meats over a low heat, combined with salting, is another effective preservation method. **Jerky** and **smoked fish** can be stored for long periods and provide a lightweight, protein-packed food source.

6.4 - Meal Planning and Rationing

In a long-term bug-in scenario, you'll need to carefully manage your food supply to avoid running out before the crisis ends. Navy SEALs are experts at rationing food and maintaining energy with minimal resources, and the same principles can be applied to your situation.

Meal Planning for Efficiency

To ensure you're making the most of your food supply, create a **meal plan** that is calorie-efficient and balanced. Plan meals that use a combination of **carbohydrates, proteins, and fats** to provide energy and sustenance without over-consuming your stockpile.

- **Batch Cooking**: Prepare meals in large batches to minimize waste and ensure that leftovers can be used for future meals. Soups, stews, and casseroles are great for batch cooking because they can be stretched over multiple days.
- **Rotating Ingredients**: Rotate ingredients with shorter shelf lives, like canned goods and dried fruits, so they are consumed before going bad. Use freeze-dried or MRE meals as a backup when other food items run low.

Rationing Your Food Supply

If the crisis lasts longer than anticipated, you'll need to implement a **rationing system** to ensure your food lasts. SEALs often have to ration food during long missions, and the following strategies can help you manage your food supply during a bug-in:

- **Portion Control**: Carefully measure out portions to avoid overeating. Each person in your household should receive a designated amount of food per meal based on their caloric needs.
- **Stretching Ingredients**: Use fillers like **rice, beans, and potatoes** to stretch meals and make your protein sources last longer. These foods are calorie-dense and inexpensive, making them ideal for rationing.
- **Eating in Stages**: Start with the most **perishable items** and consume those first. Once you've exhausted your perishable food supply,

move on to canned, dried, and freeze-dried goods, followed by your MREs or emergency food rations.

6.5 - Cooking Without Electricity

In a bug-in scenario where electricity is unavailable, you'll need to rely on **alternative cooking methods**. SEALs are trained to cook using minimal resources, and you can adapt these same methods at home.

Propane and Butane Stoves

Propane and butane camp stoves are excellent for cooking without electricity. They are portable, easy to use, and provide a reliable source of heat for boiling water, frying, or simmering meals.

- **Safety Considerations**: Always use camp stoves in a **well-ventilated area** to avoid carbon monoxide poisoning. If possible, cook outdoors or near an open window.

Wood-Burning Stoves and Fireplaces

If you have a **wood-burning stove** or fireplace, you can use it to cook during a power outage. A wood-burning stove is ideal for boiling water, simmering stews, or baking bread. You'll need a steady supply of dry wood to fuel the fire.

Solar Cookers

In sunny climates, a **solar cooker** can be a game-changer for cooking during a bug-in. Solar cookers use the sun's energy to heat and cook food, and they require no fuel or electricity.

- **How Solar Cookers Work**: Solar cookers use reflective panels to concentrate sunlight into a cooking chamber. While cooking times are longer than with traditional methods, solar cookers can reach temperatures high enough to cook meals, boil water, and bake.

6.6 - Final Thoughts: Nutrition and Survival

Securing a long-term food supply and learning how to manage it efficiently is one of the most critical aspects of surviving a bug-in scenario. Like Navy SEALs, who rely on discipline, preparation, and resourcefulness in the field, you will need to apply the same mindset to food storage and meal preparation during a crisis.

By stockpiling the right types of food, using proper preservation methods, and learning to ration your resources, you can ensure your family has enough to eat for the duration of the bug-in. Remember, survival is about more than just stockpiling; it's about using what you have wisely and being prepared to adapt as conditions change.

Chapter 7: Medical Preparedness: Self-Care and First Aid Training for Emergencies

In a crisis situation, access to professional medical help may be limited, delayed, or unavailable altogether. During a bug-in, you might find yourself or a family member in need of medical attention, and knowing how to handle common injuries, illnesses, and emergencies can be the difference between life and death. Navy SEALs are trained extensively in **field medicine**, allowing them to treat injuries, stabilize conditions, and save lives in environments where professional help isn't readily available. You must adopt a similar mindset and skill set to prepare for a long-term crisis in which emergency services may not arrive in time.

In this chapter, we'll dive into the essentials of medical preparedness. We will cover how to create a comprehensive first aid kit, the basic medical skills everyone in your household should know, and how to handle a variety of emergency medical situations. Medical preparedness isn't just about having the right tools and supplies; it's about developing the knowledge and skills to manage medical emergencies confidently, no matter the circumstances.

7.1 - Building a Comprehensive First Aid Kit: Essential Medical Supplies

One of the most important steps in medical preparedness is building a **well-stocked first aid kit**. SEALs carry comprehensive medical kits into the field to handle everything from minor cuts and burns to life-threatening trauma. Your first aid kit should be capable of addressing a wide range of injuries and illnesses that could occur during a long-term bug-in, where access to hospitals and pharmacies may be limited.

Basic First Aid Supplies

Your first aid kit should include supplies to treat common injuries such as cuts, scrapes, burns, and sprains. These injuries may not be life-threatening, but they can become serious if not treated properly, especially in a situation where medical help is unavailable.

- **Adhesive Bandages**: Include a variety of sizes of adhesive bandages (also known as **Band-Aids**) to cover small cuts and abrasions. Look for **waterproof** or **heavy-duty bandages** that stay in place during physical activity.
- **Gauze Pads and Rolls**: **Sterile gauze pads** and **rolls** are essential for covering larger wounds. They can be used to stop bleeding, protect wounds from infection, and provide padding for burns or fractures.

- **Antiseptic Wipes and Ointments**: Keeping wounds clean is crucial in a crisis, where infections can quickly become serious. Stock up on **antiseptic wipes** or solutions such as **hydrogen peroxide** or **betadine** to disinfect wounds. Include **antibiotic ointments** like **Neosporin** to apply after cleaning.
- **Medical Tape**: **Adhesive medical tape** is essential for securing bandages and gauze in place. Look for durable, waterproof tape that won't irritate the skin.
- **Tweezers and Scissors**: A high-quality pair of **tweezers** and **trauma shears** are essential for removing splinters, ticks, or debris from wounds. Trauma shears can be used to cut clothing away from an injury or to trim bandages and gauze.
- **Burn Gel or Cream**: Burns can happen easily during a crisis, especially if you're cooking over open flames or using alternative heating methods. **Burn gel** or **aloe vera gel** can soothe the pain and speed healing.
- **Instant Cold Packs**: Keep several **instant cold packs** in your first aid kit to treat sprains, strains, or bruising. These packs can be activated by squeezing or shaking and provide immediate cold therapy to reduce swelling.
- **Elastic Bandages**: **ACE bandages** or other **elastic wraps** are crucial for immobilizing sprained joints or supporting weak muscles. They can also be used to secure splints for broken bones.

- **Sterile Gloves**: Keep **disposable gloves** in your kit to protect both yourself and the patient from contamination while treating wounds. Nitrile gloves are preferable for those with latex allergies.
- **Eye Wash and Eye Pads**: Eye injuries can happen during a crisis, especially if you're dealing with smoke, debris, or chemicals. Include a **saline solution** for flushing the eyes, as well as **sterile eye pads** to protect injured eyes.

Advanced First Aid Supplies

In addition to basic supplies, your first aid kit should be equipped with more advanced items to handle **serious injuries** that require immediate attention before professional medical help can arrive.

- **Tourniquet**: A **tourniquet** is essential for controlling severe bleeding from a limb. SEALs are trained to apply tourniquets in the field, and you should learn this skill as well. Make sure to include a **CAT (Combat Application Tourniquet)** in your kit, as these are widely used in military settings.
- **Hemostatic Dressings**: For severe wounds, **hemostatic dressings** like **QuikClot** or **Celox** can help stop heavy bleeding by promoting rapid blood clotting. These dressings are used in combat to treat traumatic injuries and can be lifesaving in a bug-in scenario.

- **SAM Splint**: The **SAM Splint** is a lightweight, flexible splint that can be molded to stabilize broken bones or sprains. It's an essential tool for immobilizing fractures, and its compact design makes it easy to store in your kit.
- **CPR Mask or Shield**: A **CPR mask** or **shield** is crucial for protecting both you and the patient while performing **cardiopulmonary resuscitation (CPR)**. This barrier prevents the spread of germs while providing life-saving breaths.
- **Trauma Dressings**: Large **trauma dressings** are used to cover serious wounds, such as those caused by deep cuts, gunshot wounds, or severe burns. These dressings are highly absorbent and provide pressure to stop bleeding.
- **Suture Kit or Steri-Strips**: For deep cuts that may require stitching, having a **suture kit** or **Steri-Strips** (butterfly bandages) can help close the wound temporarily until professional medical care is available. However, stitching a wound is a complex task that requires training, so it's important to know how to use these tools properly.
- **Chest Seal**: A **chest seal** is used to treat a **sucking chest wound**, which occurs when air enters the chest cavity, typically due to a puncture or gunshot. This can lead to a collapsed lung, and the chest seal helps prevent air from entering the wound.

Medications and Personal Health Supplies

In addition to first aid supplies, it's critical to include **medications** and **personal health items** in your medical preparedness kit.

- **Pain Relievers**: Stock up on **over-the-counter pain relievers** such as **ibuprofen**, **acetaminophen**, and **aspirin** to manage pain, reduce inflammation, and lower fevers.
- **Antihistamines**: Include **antihistamines** such as **Benadryl** or **Zyrtec** to treat allergic reactions, bug bites, or other skin irritations.
- **Anti-Diarrheal Medications**: Dehydration caused by diarrhea can be life-threatening in a crisis. **Anti-diarrheal medications** such as **Imodium** should be included in your kit.
- **Prescription Medications**: If anyone in your household takes **prescription medications**, ensure you have a **supply that will last at least one to two months**. Speak with your doctor about obtaining extra medication for emergencies.
- **Electrolyte Tablets or Powder**: During illness or extreme physical exertion, **electrolytes** are lost through sweat and bodily fluids. Keep a supply of **electrolyte tablets** or **powder** to replenish lost minerals and prevent dehydration.

7.2 - First Aid Skills: Basic Training Everyone Should Know

Even the most well-stocked first aid kit is useless without the knowledge of how to use it. In the military, Navy SEALs undergo extensive first aid and trauma training to ensure that they can respond to medical emergencies in the field. While you won't need the level of training SEALs receive, knowing **basic first aid skills** is essential for every member of your household.

Here are the core first aid skills everyone should be familiar with:

How to Perform CPR

CPR (Cardiopulmonary Resuscitation) is a life-saving technique used in cases of cardiac arrest, where the heart has stopped beating. Knowing how to perform CPR can significantly increase a person's chances of survival until professional help arrives.

- **Chest Compressions**: Place the heel of your hand in the center of the patient's chest, just above the sternum. Place your other hand on top, interlock your fingers, and keep your arms straight. Press down hard and fast, aiming for a rate of **100-120 compressions per minute**. Push down **2 inches** deep in adults and release completely after each compression.

- **Rescue Breaths**: After 30 compressions, give **two rescue breaths** by tilting the person's head back, pinching their nose, and breathing into their mouth until you see their chest rise.
- **Repeat**: Continue alternating between 30 chest compressions and 2 rescue breaths until professional help arrives or the person begins breathing.

How to Control Bleeding

Controlling **bleeding** is one of the most important first aid skills you can learn, especially in a situation where medical help is delayed. Whether from a deep cut, accident, or trauma, severe bleeding can lead to shock or death if not treated promptly.

- **Direct Pressure**: Apply **direct pressure** to the wound using a clean cloth or sterile gauze. Hold the pressure for several minutes without lifting the cloth to check the wound.
- **Elevation**: If possible, **elevate the injured area** above the level of the heart to reduce blood flow.
- **Tourniquet Use**: If the bleeding is from an arm or leg and cannot be controlled with direct pressure, apply a **tourniquet**. Place the tourniquet **2-3 inches above the wound**, tighten it until the bleeding stops, and secure it in place. Note the time the tourniquet was applied, as leaving it on for too long can cause tissue damage.

How to Treat Shock

Shock occurs when the body's organs don't receive enough oxygenated blood, which can happen after significant blood loss, severe infection, or trauma. It's a life-threatening condition that requires immediate attention.

- **Lay the Person Down**: Have the person **lie flat on their back** and elevate their legs about **12 inches** to help improve circulation.
- **Keep Them Warm**: Use a **blanket** or **jacket** to keep the person warm. Shock can cause body temperature to drop rapidly.
- **Monitor Breathing**: Keep an eye on the person's breathing and **check their pulse regularly**. If they stop breathing, be ready to perform CPR.
- **Do Not Give Food or Water**: Do not allow the person to eat or drink, as it can worsen the condition. Focus on keeping them calm and still until help arrives.

How to Splint a Broken Bone

In a crisis, **broken bones** or fractures are common injuries that must be stabilized to prevent further damage. Knowing how to apply a **splint** will help keep the bone in place until professional medical care is available.

- **Immobilize the Area**: Use a **rigid object** like a wooden stick, rolled-up newspaper, or SAM splint to immobilize the broken bone. Place the splint so it extends above and below the injured area.
- **Secure the Splint**: Use **bandages, belts, or strips of cloth** to secure the splint to the injured limb. Tie the splint tightly enough to hold the bone in place, but not so tight that it cuts off circulation.
- **Elevate and Ice**: If possible, **elevate the injured limb** and apply an **ice pack** to reduce swelling.

7.3 - Medical Conditions to Be Aware Of: Illnesses and Infections

In a long-term bug-in scenario, illnesses and infections can become more common due to limited access to medical care, poor hygiene, and compromised immune systems. Navy SEALs are trained to recognize the signs of common illnesses and infections in the field, and it's important for you to know how to identify and treat them in a bug-in situation.

Dehydration

Dehydration is a serious condition that can occur when the body loses more fluids than it takes in. It can happen quickly during hot weather, illness, or physical exertion, and it's especially dangerous for children and the elderly.

- **Symptoms**: Look for signs of dehydration such as **dry mouth, dizziness, dark urine, and lethargy**. Severe dehydration can cause **confusion, fainting**, and **rapid heart rate**.
- **Treatment**: Rehydrate by drinking **small sips of water** or an **oral rehydration solution** that contains electrolytes. If the person is unable to drink, apply **wet cloths** to their skin to help cool them down.

Infections and Sepsis

Without access to antibiotics, even minor cuts and scrapes can lead to infections. Left untreated, these infections can develop into **sepsis**, a life-threatening condition in which the body's immune system attacks its own tissues.

- **Signs of Infection**: Look for **redness, swelling, heat, and pus** around a wound. If the person develops a **fever, chills**, or **rapid breathing**, the infection may be spreading.
- **Treatment**: Clean the wound thoroughly with **antiseptic** and apply **antibiotic ointment**. Keep the area covered with sterile bandages and monitor for signs of improvement. If the infection worsens, you may need to administer **oral antibiotics** if available.

Heat Exhaustion and Heatstroke

Prolonged exposure to high temperatures can lead to **heat exhaustion** or **heatstroke**, both of which are serious conditions that require immediate treatment.

- **Symptoms of Heat Exhaustion**: Look for **heavy sweating, weakness, nausea, and dizziness**. If untreated, heat exhaustion can progress to heatstroke.
- **Symptoms of Heatstroke**: **Heatstroke** is a medical emergency and occurs when the body's temperature rises above 104°F (40°C). Symptoms include **confusion, rapid heart rate, headache**, and **unconsciousness**.
- **Treatment**: Move the person to a **cool, shaded area** and remove excess clothing. Apply **cold compresses** or ice packs to the neck, armpits, and groin, and have the person sip cool water. In the case of heatstroke, seek medical help immediately.

Hypothermia

In cold weather conditions, hypothermia can set in when the body's temperature drops below **95°F (35°C)**. It's a life-threatening condition that requires immediate warming to prevent organ failure.

- **Symptoms**: Look for **shivering, confusion, slurred speech, and clumsiness**. As

hypothermia progresses, the person may stop shivering and become lethargic or unresponsive.
- **Treatment**: Move the person to a **warm, dry environment**. Replace wet clothing with dry layers, and wrap the person in blankets. Offer **warm liquids** if the person is conscious, but avoid alcohol or caffeine.

7.4 - Creating a Medical Emergency Plan

Just as SEALs create detailed plans for every mission, you'll need a **medical emergency plan** to ensure that everyone in your household knows how to respond to injuries and illnesses during a bug-in. Your medical plan should include the following:

- **Roles and Responsibilities**: Assign roles to each family member based on their skills and abilities. Designate someone as the primary **first responder**, while others can assist with gathering supplies, monitoring the patient, or providing emotional support.
- **Emergency Contacts**: Write down emergency contact numbers, including those for local hospitals, doctors, or trusted neighbors who have medical experience. If phone lines are down, ensure that you have alternative ways to reach out for help, such as **two-way radios** or **satellite phones**.
- **First Aid Drills**: Practice **first aid drills** regularly with your household. Everyone should know how

to perform CPR, apply bandages, and use a tourniquet in case of an emergency.
- **Medical Logs**: Keep a **medical logbook** where you can record any injuries, illnesses, and treatments. This is especially important in long-term crises, as you may need to track the progression of a condition or remember the timing of medication dosages.

7.5 - Final Thoughts: Becoming Your Own First Responder

Medical preparedness is one of the most critical aspects of surviving a long-term bug-in scenario. By learning **first aid skills**, stocking up on essential **medical supplies**, and creating a detailed **emergency plan**, you'll be ready to handle medical emergencies when professional help is unavailable.

Remember, SEALs are trained to be their own first responders, and with the right preparation, you can do the same. Medical knowledge isn't just about responding to emergencies; it's about **preventing injuries** and maintaining the health and well-being of your household throughout the crisis. The more prepared you are, the greater your chances of surviving and thriving in any situation.

Chapter 8: Communication and Intelligence: Staying Informed and Connected

During a crisis, especially one that leads to a long-term bug-in, reliable communication is one of the most crucial aspects of survival. Without access to information or the ability to connect with others, you're left vulnerable to misinformation, isolation, and an inability to make informed decisions. Navy SEALs operate in some of the world's most dangerous environments, often cut off from standard lines of communication. Despite these challenges, they remain connected to their teams and gather intelligence to make strategic decisions.

In this chapter, we will explore the various methods of maintaining communication during a bug-in scenario, how to gather critical intelligence, and how to ensure your home remains a stronghold of information and awareness. Whether it's monitoring emergency broadcasts, maintaining contact with family members, or staying informed about the situation outside, having a solid plan for communication is vital to your survival.

8.1 - Why Communication Is Critical in a Crisis

In any crisis, communication is the **lifeline** to the outside world. It allows you to stay updated on the evolving situation, whether it's the status of a natural disaster, a pandemic, or civil unrest. Navy SEALs are trained to constantly gather and interpret **real-time intelligence** to adjust their strategies and make decisions based on the most accurate information available.

For you, communication serves several purposes:

- **Situational Awareness**: Knowing what's happening beyond your immediate surroundings is crucial for making informed decisions. Whether it's a weather update, news of nearby dangers, or directives from local authorities, situational awareness keeps you ahead of potential threats.
- **Coordination**: If you're bugging in with family or neighbors, you'll need a reliable way to communicate to coordinate efforts, plan responses, and share resources.
- **Emergency Response**: In the event of an emergency, such as a medical situation or home intrusion, being able to contact others for help can save lives.
- **Psychological Health**: Staying connected to others, whether through local networks or broader communications, helps combat the **psychological stress** of isolation.

8.2 - Emergency Communication Methods: Staying Connected Without the Grid

When power is down, cell towers are overloaded, or the internet becomes unreliable, you need alternative methods of communication to stay connected. SEALs are trained to use a wide range of communication devices in the field, from radios to satellite systems, and similar tools can be applied during a bug-in.

AM/FM and NOAA Radios

A **battery-powered** or **hand-crank AM/FM radio** is one of the most basic yet essential communication tools you should have. In a bug-in scenario, especially when the power grid is down, these radios allow you to receive critical updates from **local authorities**, **emergency services**, and the **National Oceanic and Atmospheric Administration (NOAA)** weather alerts.

- **AM/FM Radios**: AM stations often provide the widest range of emergency broadcasts, including **public safety updates** and **evacuation orders**. FM stations tend to focus on **local news** and can keep you informed about the specific conditions in your area.
- **NOAA Weather Radios**: NOAA broadcasts 24/7 weather updates, natural disaster warnings, and other public safety messages. Having a **dedicated NOAA weather radio** is essential if

you're in an area prone to hurricanes, tornadoes, floods, or wildfires.

Hand-Crank and Solar-Powered Radios

To ensure that your radio remains functional when batteries run out, invest in **hand-crank** or **solar-powered radios**. These radios do not rely on external power sources, making them ideal for long-term bug-ins. Many models also come with built-in **flashlights** and **USB charging ports**, allowing you to power small devices like phones.

- **Hand-Crank Radios**: These radios have a manual crank that generates power to keep the radio operational. Even if you're without sunlight or batteries, you can crank the radio to receive broadcasts.
- **Solar-Powered Radios**: Solar-powered radios use the sun's energy to charge. This is a more passive method of powering your communication devices and can be extremely useful if you're in a sunny environment with limited battery access.

Two-Way Radios (Walkie-Talkies)

Two-way radios, or **walkie-talkies**, are crucial for **short-range communication** within your household or between neighbors. If the power grid and cell networks are down, these devices allow you to communicate across distances of **1 to 5 miles**, depending on the terrain and the quality of the radio.

- **General Mobile Radio Service (GMRS)**: GMRS radios offer more range and power than standard walkie-talkies, with coverage of **up to 10-15 miles** in open areas. They require a **license** to operate but are invaluable for keeping in contact with family members or nearby groups.
- **Family Radio Service (FRS)**: FRS radios are lower-power walkie-talkies that do not require a license. They typically cover a shorter range (around 2 miles) but are useful for intra-household communication during a bug-in.
- **Emergency Channels**: Many two-way radios come pre-programmed with **emergency channels**, allowing you to broadcast distress signals or listen for updates from others in your community.

CB Radios (Citizens Band Radios)

CB radios are a long-established means of communication, especially in rural areas or among truckers. They are useful during a crisis because they allow you to **communicate over longer distances** and **monitor public channels**.

- **Range**: The range of a CB radio typically falls between **3 to 20 miles**, depending on the terrain and the power of the radio. While not as powerful as GMRS, they provide a reliable way to stay connected in local or regional crises.

- **Channel 9**: **Channel 9** is designated as the **emergency channel** on CB radios, and it's monitored by local authorities, especially in rural areas. In case of an emergency, this channel can be used to call for help or broadcast distress signals.

Ham Radios (Amateur Radios)

Ham radios, or amateur radios, are one of the most powerful communication tools available during a crisis. Unlike other communication devices, ham radios offer **long-distance communication** and can even **reach across continents**. SEALs sometimes rely on similar systems for secure, long-range communication.

- **Licensing**: Operating a ham radio in the United States requires a **license** from the Federal Communications Commission (FCC). However, during an emergency, unlicensed use is permitted to call for help.
- **Capabilities**: Ham radios allow you to communicate over hundreds of miles, giving you access to a wide network of amateur radio operators who often act as a critical lifeline during disasters. Many ham radio operators are part of **emergency response networks** and can provide real-time information about the situation on the ground.

Satellite Phones

In the most severe crises, when all other communication methods fail, a **satellite phone** provides an independent, global communication option. SEAL teams use satellite communications for remote operations, and while more expensive than other devices, satellite phones offer **uninterrupted communication** anywhere in the world.

- **How It Works**: Unlike cell phones, which rely on ground-based towers, satellite phones communicate directly with orbiting satellites. This allows for clear communication even in remote or disaster-stricken areas.
- **Cost and Accessibility**: While satellite phones are a significant investment, they offer the most reliable communication in an extended crisis. Some models also allow for **text messaging** and **data transmission**, making them versatile tools for staying connected.

8.3 - Communication Security: Protecting Your Information

Maintaining **communication security** is critical during a crisis, especially if the situation involves civil unrest, lawlessness, or looting. SEALs are trained to operate under conditions where secure communication is paramount. In a long-term bug-in, you must also protect your **private information** and ensure that your communications aren't intercepted or compromised.

Encryption and Secure Channels

If you are using **two-way radios**, **CB radios**, or **ham radios**, it's important to understand that these devices operate on **open channels**. This means that anyone with a similar device can listen in on your communications, potentially compromising your security.

- **Code Words**: Use **code words** or **phrases** to communicate sensitive information without revealing specifics to potential eavesdroppers. SEAL teams often use code names for locations, actions, or objectives, and you can create similar systems for your family.
- **Scramblers**: Some advanced two-way radios and communication systems come with **scrambling technology**, which makes it more difficult for outsiders to listen to your conversations. If possible, invest in a system with built-in scrambling or encryption features.

Operational Security (OPSEC)

In military operations, **OPSEC (Operational Security)** refers to the practice of safeguarding sensitive information to prevent adversaries from gaining a tactical advantage. During a bug-in, maintaining OPSEC is just as important. Be mindful of what information you're broadcasting and who might be listening.

- **Don't Reveal Too Much**: Avoid discussing your supplies, weapons, or security measures over

open communication channels. Broadcasting this information can make you a target for looters or other opportunists.
- **Be Vague About Locations**: When coordinating with others, avoid giving exact addresses or locations. Instead, use general landmarks or pre-arranged meeting points.
- **Limit Unnecessary Communications**: SEALs maintain a strict communication discipline, speaking only when necessary to avoid giving away their position or plans. Apply this same principle in a crisis—avoid idle chatter over open channels.

8.4 - Gathering Intelligence: Staying Informed During a Crisis

In a crisis, intelligence isn't just about gathering news—it's about obtaining actionable information that allows you to make informed decisions. SEALs are experts in **intelligence gathering**, using all available resources to understand the situation and adjust their strategies. During a bug-in, you must gather and analyze information to stay ahead of evolving threats.

Monitoring Local News and Emergency Broadcasts

In the early stages of a crisis, local news outlets and emergency broadcasters will be your primary sources of information. Use your **AM/FM radio** to listen to **local**

updates, evacuation orders, and news about public safety efforts.

- **Prioritize Emergency Alerts**: Focus on government-issued **emergency alerts**, such as those provided by FEMA, NOAA, or your local emergency management office. These updates often include information about **power outages**, **curfews**, **emergency shelters**, and other critical details.
- **Weather Updates**: If the crisis involves natural disasters, such as hurricanes, tornadoes, or wildfires, listen for **weather updates** and forecasts. Knowing the path of a storm or the spread of a wildfire can help you decide whether to stay put or evacuate.

Intelligence from Neighbors and Local Networks

In a bug-in scenario, your neighbors can be an invaluable source of intelligence. SEALs often rely on **local informants** to understand the dynamics of a region, and you can create your own local network to share resources and information.

- **Establish a Communication Network**: Set up a communication network with your immediate neighbors, using **two-way radios** or **walkie-talkies** to keep in touch. This allows you to exchange updates on local conditions, such

as nearby threats, availability of supplies, or the status of emergency services.
- **Coordinate Patrols or Watch Shifts**: In cases of civil unrest or looting, organize **neighborhood patrols** or **watch shifts** to monitor the area and alert each other to potential threats. Having a few trusted neighbors working together can greatly enhance your situational awareness and security.

Online Sources and Social Media

If the internet is still operational, **online sources** and **social media platforms** can provide real-time updates. However, SEALs are trained to filter through misinformation, and you should apply the same level of caution when using these sources.

- **Follow Official Accounts**: Follow **official government accounts**, local authorities, and reputable news organizations for reliable updates. Avoid unverified or sensational posts that could spread false information.
- **Check Multiple Sources**: To avoid falling victim to **misinformation**, cross-reference the news you receive with other trusted sources. SEALs rely on **multiple intelligence channels** to verify the accuracy of information, and you should do the same.

8.5 - Psychological Preparedness: Staying Calm in Isolation

Prolonged isolation can take a significant toll on your mental health. Navy SEALs are trained to endure long periods of isolation, often without communication or contact with the outside world. During a bug-in, it's essential to prioritize **psychological preparedness** to avoid feelings of despair, anxiety, or depression.

Combatting Isolation and Loneliness

Isolation can cause **cabin fever** and feelings of loneliness, especially if communication with the outside world is cut off. Here's how SEALs stay mentally resilient in isolated environments, and how you can apply those techniques during a bug-in.

- **Maintain Routines**: Routines create a sense of normalcy, even in chaotic situations. Establish daily routines for **meals, exercise, chores**, and **communication**, which will help structure your day and keep you grounded.
- **Stay Engaged**: Keep your mind engaged by learning new skills, reading, or teaching others. SEALs often use downtime to **study** or **practice**, and doing the same will keep your mind sharp.
- **Connect with Loved Ones**: If possible, maintain contact with family members, friends, or neighbors. Whether through phone calls, two-way radios, or in-person conversations,

staying connected with others is crucial for **emotional support**.

Stress Management and Mental Resilience

High-stress situations, such as long-term crises, can lead to **panic**, **anxiety**, and **mental fatigue**. SEALs use several techniques to manage stress and stay focused in high-pressure environments.

- **Deep Breathing and Meditation**: Practicing **deep breathing techniques** and **meditation** can help reduce stress and anxiety. Simple breathing exercises like the **4-7-8 technique** (inhale for 4 seconds, hold for 7 seconds, exhale for 8 seconds) can calm your nervous system and clear your mind.
- **Positive Visualization**: SEALs often use **visualization techniques** to mentally prepare for missions. Visualizing successful outcomes can help you stay motivated and focused, even when the situation seems dire.

8.6 - Final Thoughts: Communication as a Lifeline

In a crisis, communication is far more than just staying in touch—it's your lifeline to the outside world, a source of critical information, and a way to maintain your psychological well-being. SEALs are taught that communication can make or break a mission, and in a long-term bug-in scenario, your ability to stay connected and informed will be key to your survival.

By preparing the right tools, understanding how to secure your communications, and gathering actionable intelligence, you can ensure that your household remains aware, connected, and safe, no matter what the crisis throws at you.

Chapter 9: Defending Your Home: Security Tactics for Bugging In

In a long-term crisis scenario, one of your greatest concerns will be **securing your home** against external threats. Whether it's opportunistic looters, civil unrest, or desperate individuals seeking supplies, your home could become a target if the situation outside deteriorates. Navy SEALs are trained in both **offensive and defensive tactics**, often operating in hostile environments where they need to secure bases, camps, and positions from attack. Applying similar **defensive strategies** will help you turn your home into a stronghold capable of withstanding external threats.

While the focus of this guide has been on **bugging in**, remaining inside your home for safety, that safety will only be as strong as your **defensive measures**. You must be proactive in fortifying your perimeter, securing entry points, and developing a defense plan that not only deters potential threats but also protects you and your family should an intruder breach your home.

In this chapter, we'll cover how to establish layers of defense around your home, how to reinforce entry points, and how to implement tactical strategies for **home defense** based on **Navy SEAL training**.

9.1 - Establishing Layers of Defense: Creating a Secure Perimeter

In military operations, SEALs are trained to establish **multiple layers of defense** when securing a base or strategic position. These layers are designed to detect and repel threats at various stages before they can penetrate the core of the defensive stronghold. You can use the same principles to secure your home, creating **concentric rings of security** that make it increasingly difficult for intruders to reach your house.

The First Layer: Perimeter Security

Your **perimeter** is the first line of defense against external threats. Whether you live in a house with a yard, an apartment with a small entryway, or a remote property, the area surrounding your home must be made as secure as possible to prevent unwanted intrusions.

Fencing and Gates

One of the most effective ways to secure your perimeter is with a strong **fence**. While a basic fence may provide some privacy, you'll need to take extra steps to ensure that it offers real protection against intruders.

- **Height**: A fence that is at least **6-8 feet tall** can deter casual intruders or looters looking for easy targets. Consider adding **barbed wire** or **thorny**

- **plants** along the top to make scaling it more difficult.
- **Material**: **Chain-link** or **wrought iron fences** allow for visibility, preventing intruders from hiding behind them while trying to breach your property. On the other hand, **wooden or vinyl fences** offer more privacy but may require reinforcement to prevent break-ins.
- **Gates**: Any gate on your property should be as secure as your fencing. Use **high-quality locks** or **padlocks** to secure gates, and consider adding **security bars** or **grills** to prevent forced entry.

Lighting and Cameras

Lighting is one of the simplest but most effective deterrents against intruders. Darkness offers cover to those attempting to breach your perimeter, so eliminating dark corners and shadowy areas will reduce the likelihood of a break-in.

- **Motion-Sensing Lights**: Install **motion-activated lights** around the perimeter of your property. Place them near **entry points**, **driveways**, and **dark corners**. These lights will turn on when movement is detected, startling potential intruders and alerting you to their presence.
- **Floodlights**: For larger properties or outdoor areas, **floodlights** can be used to illuminate a

wide area. These lights should be placed high up to cover large swaths of your yard or driveway.
- **Security Cameras**: SEALs rely on **reconnaissance** to gather intelligence, and security cameras act as your eyes and ears on your perimeter. Install **outdoor security cameras** that cover the **front, back, and sides** of your property. Make sure they are high enough to avoid tampering and that they have **night vision** capabilities. If possible, integrate the cameras with **motion alerts** so you'll be notified of any movement on your property.

The Second Layer: Yard or Entryway Defense

Once your perimeter is secure, the next layer of defense is the area immediately surrounding your home, whether it's a yard, driveway, or entryway. This space should be optimized to slow down and expose intruders, making it difficult for them to reach your house undetected.

Defensive Landscaping

Landscaping can be a powerful defensive tool when done correctly. SEALs are trained to use the environment to their advantage, and you can do the same with your yard or property.

- **Thorny Plants**: Plant **thorny bushes** or **brambles** beneath windows and along the perimeter of your house. Plants like **rose bushes**, **holly**, or **blackberry vines** create a

natural barrier that is difficult to navigate without injury.
- **Gravel Paths**: Consider laying **gravel or stone paths** around your house, particularly near entry points. Gravel creates noise when walked on, alerting you to any movement outside your home.
- **Clearing Vegetation**: Keep trees, shrubs, and other vegetation **trimmed back** away from windows and doors. Dense vegetation provides cover for intruders attempting to approach your house undetected.

Obstacles and Barriers

Physical obstacles can slow down intruders and make it harder for them to reach your home. SEALs are taught to **channel** potential threats into areas where they can be neutralized, and you can apply the same principle to your property.

- **Driveway Gates**: Install a **heavy-duty gate** at the entrance of your driveway to control vehicle access. Use **padlocks** or **electronic locks** to secure the gate.
- **Barricades**: Use **heavy planters, large rocks, or concrete barriers** to block vehicle access to certain parts of your property. These obstacles make it more difficult for vehicles to drive up close to your house, reducing the risk of a forced entry by vehicle.

9.2 - Reinforcing Entry Points: Securing Doors, Windows, and Access Points

The **entry points** to your home—your doors, windows, and garage—are the most vulnerable areas when it comes to a potential break-in. Once an intruder has made it past your perimeter and yard, these are the final barriers that stand between them and your family. Navy SEALs are experts at **breaching** and **securing** entry points, and knowing how to reinforce your home's vulnerabilities is critical to your defense strategy.

Reinforcing Doors

Your **doors** are the primary access points to your home, and they should be **fortified** to withstand forced entry attempts. A basic door with a standard lock isn't enough to keep determined intruders out.

Solid-Core or Metal Doors

Start by ensuring that all of your exterior doors are **solid-core** or made of **reinforced steel**. Hollow-core doors are easy to kick in and provide little resistance to brute force. **Metal doors** offer greater security and are difficult to breach without specialized tools.

- **Reinforced Door Frames**: Even the strongest door can be compromised if the door frame is weak. Reinforce your door frames with **metal**

strike plates and **long screws** that anchor into the wall studs.
- **Deadbolt Locks**: Install **high-quality deadbolts** on all exterior doors. The deadbolt should extend at least **1 inch** into the door frame. For even more security, opt for **double-cylinder deadbolts**, which require a key to unlock from both sides. This prevents intruders from breaking a nearby window and unlocking the door from the inside.
- **Door Barricades**: Consider installing a **door barricade** or **security bar** that can be placed across the inside of the door to prevent forced entry. These devices offer an extra layer of protection, particularly for **front and back doors**.

Security Doors

For an added layer of defense, install a **security door** in front of your main entrance. These doors are made of **reinforced steel or wrought iron** and provide a secondary barrier to anyone attempting to break in. They also allow you to answer the door without fully exposing yourself to potential threats.

Securing Windows

Windows are often the weakest point of entry in any home. SEALs are trained to exploit vulnerabilities like windows when breaching a target, and intruders will do the same if your windows are not properly secured.

Window Locks and Bars

Start by ensuring that all of your windows are equipped with **high-quality locks**. Standard window latches are easy to break or bypass, so consider upgrading to **keyed locks** or **pin locks** that prevent the window from being opened from the outside.

- **Window Bars**: Install **window security bars** on **ground-floor windows** or any windows that are easily accessible from the outside. These bars make it nearly impossible for an intruder to gain entry through a window. If aesthetics are a concern, you can opt for **decorative bars** that provide security without compromising the appearance of your home.
- **Window Alarms**: Install **battery-operated window alarms** that trigger a loud alert when a window is opened or broken. These alarms are inexpensive and can deter intruders by drawing attention to their attempts to break in.

Reinforced Glass and Window Film

For added protection, consider installing **shatter-resistant window film** or **laminated security glass** on all windows. These products make it significantly more difficult for an intruder to break through the glass, even with tools.

- **Shatterproof Film**: Security window film is a clear adhesive that can be applied to the inside

of your windows. While it won't prevent the glass from breaking, it holds the shattered glass in place, making it harder for an intruder to create a large enough opening to enter.
- **Laminated Security Glass**: For maximum protection, replace standard glass windows with **laminated security glass**. This glass consists of multiple layers that are bonded together, creating a barrier that is resistant to impacts and shattering.

Securing the Garage

Your **garage** is another vulnerable access point that intruders can exploit to gain entry to your home. Many homeowners overlook garage security, but SEALs know that every access point must be secured to prevent a breach.

Garage Door Locks and Bars

Most garage doors rely on an **electric opener** for security, but these can be bypassed if the power is out or if the intruder has the right tools. To prevent this, install a **manual garage door lock** that secures the door from the inside. You can also place a **garage door security bar** across the inside of the door to block it from being opened.

Disable the Emergency Release

One common trick used by intruders is to trigger the **emergency release cord** on automatic garage doors, which allows them to open the door manually. To prevent this, disable the emergency release cord when you're bugging in, or secure it with a **zip tie** to prevent tampering.

9.3 - Tactical Home Defense: Preparing for Worst-Case Scenarios

While the goal is to prevent intruders from breaching your home, you must also be prepared for worst-case scenarios in which an intruder makes it inside. SEALs are trained in **close-quarters combat (CQC)** and **defensive tactics**, which can be adapted for home defense. This section covers how to prepare for and respond to home invasions, should they occur.

Safe Rooms and Retreat Points

One of the most important aspects of home defense is having a **safe retreat point** in case an intruder makes it inside. SEALs always have an escape or fallback plan, and you should have one as well.

- **Choosing a Safe Room**: The best safe rooms are located in **interior rooms** without windows, such as a bathroom, large closet, or basement. The door to the room should be reinforced with a

solid core door and a **deadbolt**. Install a **security bar** or **door brace** for extra protection.
- **Stocking Your Safe Room**: Your safe room should be stocked with essential supplies, such as **water**, **non-perishable food**, **flashlights**, **first aid kits**, and **communication devices** (like a two-way radio or cell phone). It's also wise to keep **self-defense tools** in the room.

Self-Defense Tools

While firearms are often the preferred self-defense tool, not everyone is comfortable or trained to use them. Navy SEALs are experts in a wide range of weapons, from firearms to non-lethal options. Here's how to choose the right self-defense tools for your home:

- **Firearms**: If you choose to have a firearm for home defense, ensure that you are **properly trained** and that the weapon is stored **safely** but remains **easily accessible** in an emergency. A **shotgun** or **handgun** is often the best choice for close-quarters situations.
- **Non-Lethal Weapons**: For those who prefer non-lethal options, consider **pepper spray**, **stun guns**, or **tactical batons**. These tools can incapacitate an intruder without causing permanent harm, allowing you to escape or call for help.

Emergency Drills

Just as SEAL teams conduct regular **tactical drills**, you should practice **home defense drills** with your family. Everyone should know the plan in case of an intruder, including where to go, how to secure doors, and how to communicate with others in the household.

- **Evacuation Routes**: Identify the safest **escape routes** from each room in the house, whether through windows or doors. Make sure these routes are clear of obstacles and can be accessed quickly.
- **Silent Alarms**: Install **silent alarms** or use **panic buttons** that can alert law enforcement or trigger loud alarms without drawing attention to yourself.

9.4 - Psychological Preparedness: Staying Calm Under Threat

Defending your home isn't just about physical barriers and tools; it's also about your **mental and emotional state**. Navy SEALs are trained to remain calm and focused under extreme pressure, and you must cultivate the same mindset when facing a potential threat.

Dealing with Fear

Fear is a natural response to danger, but it can also cloud your judgment and cause you to make mistakes.

SEALs learn to channel their fear into **calculated action** by staying focused on the mission. If an intruder is attempting to breach your home, remain calm and focus on your defensive plan. Trust in the preparations you've made and follow your home defense procedures.

Situational Awareness

In a high-stress situation, maintaining **situational awareness** is critical. SEALs are trained to scan their surroundings constantly, identifying potential threats and escape routes. In your home, keep an ear out for unusual sounds, movements, or changes in lighting that may indicate an intruder. Always be aware of your surroundings, especially in areas of the house that are more vulnerable to attack.

9.5 - Final Thoughts: Securing Your Fortress

Securing your home during a long-term crisis requires **proactive planning**, **reinforcement of entry points**, and **tactical preparation** for worst-case scenarios. By establishing **layers of defense**, reinforcing **doors and windows**, and practicing **home defense drills**, you'll be ready to protect your home and family from external threats. Remember, survival isn't just about hiding—it's about creating a stronghold that can withstand any challenge that comes your way.

Navy SEALs are masters of defensive strategy, and by adopting these tactics, you can transform your home into a **fortress of safety** during any crisis.

Chapter 10: Sustainable Living: Creating Long-Term Systems for Water, Food, and Energy

In a short-term crisis, having a stockpile of supplies can get you through without significant disruptions. However, during a long-term emergency, whether due to natural disasters, civil unrest, or infrastructure collapse, those stockpiles may run out or become insufficient. At this point, your focus should shift from relying solely on stored resources to creating sustainable systems that provide **ongoing access to food, water, and energy**. In survival scenarios, Navy SEALs often have to create their own systems for water purification, food acquisition, and power generation to sustain themselves in hostile environments. Drawing on these principles, this chapter will guide you through creating a self-sufficient home capable of sustaining you and your family for the long haul.

10.1 - Establishing a Long-Term Water System

Water is essential for survival, as we discussed in **Chapter 5**, but in a long-term crisis, you may not have consistent access to potable water from traditional sources. Whether the municipal water system has failed, your well has run dry, or bottled water is no longer available, you will need to find alternative ways to secure and purify water. SEALs are trained to source and purify water from their environment, and the same methods can be adapted for use in your home.

Rainwater Harvesting

One of the most effective ways to create a sustainable water supply is by **harvesting rainwater**. Rainwater collection is a practical solution for providing water for drinking, cooking, cleaning, and irrigation, especially during an extended crisis where other sources may be unavailable.

Setting Up a Rainwater Collection System

- **Gutters and Downspouts**: Your home's existing **gutters and downspouts** can be used to funnel rainwater into a storage container. Make sure your gutters are clean and free of debris to ensure the water collected is as pure as possible.
- **Rain Barrels**: Install **rain barrels** or **large water tanks** at the base of your downspouts to collect the water. These barrels should be made of **food-grade plastic** or **stainless steel** to avoid contamination. Most rain barrels come with built-in **spigots** for easy access to the water and can be connected to a larger storage system if necessary.
- **First Flush Diverters**: When it first begins to rain, the initial water flow is likely to be

contaminated with dirt, debris, and pollutants from your roof. A **first flush diverter** is a simple device that directs this contaminated water away from your main storage tank, ensuring that only clean water enters your rain barrels.
- **Water Filtration and Purification**: Rainwater can be used directly for irrigation and cleaning, but it must be **purified** for drinking. Use a **gravity-fed water filter**, such as a **Berkey system**, or boil the water to kill pathogens. For added protection, you can also treat the water with **purification tablets** or **chlorine drops**.

Maximizing Water Storage

- **Storage Tanks**: If you live in an area with frequent rainfall, consider investing in larger **storage tanks** that can hold several hundred or even thousands of gallons of water. These tanks can be installed above or below ground, depending on space availability.
- **Water Preservation**: Keep your water storage tanks covered to prevent **algae growth** and **contamination** from insects or debris. Rotate stored water regularly, and ensure that any water used for drinking is properly filtered and purified.

Alternative Water Sources

In addition to rainwater collection, it's important to identify other **natural water sources** that can sustain your household during a long-term bug-in. SEALs are often trained to locate and use alternative water sources in the field, and you can apply similar techniques at home.

- **Nearby Streams or Rivers**: If you live near a **stream, river, or pond**, these can serve as supplementary water sources during a crisis. However, untreated water from natural sources often contains **bacteria, parasites**, and **chemical pollutants**, so purification is essential. Use a **portable water filter**, like a **LifeStraw** or **Sawyer Mini**, or boil the water before drinking.
- **Wells**: If you have a **well** on your property, consider installing a **manual pump** or **solar-powered pump** to access the water in case of a power outage. Wells can provide a reliable source of water, but regular maintenance is required to ensure the water remains safe to drink.

10.2 - Sustainable Food Production: Growing and Preserving Your Own Food

In a long-term bug-in, the food you have stored will eventually run out. When that happens, you'll need a sustainable way to produce your own food to ensure your family's continued survival. SEALs, when operating in remote environments, often rely on foraging, hunting, and fishing to supplement their rations. You can adopt similar strategies by creating a **self-sufficient food**

system at home through gardening, animal husbandry, and food preservation techniques.

Creating a Survival Garden

A **survival garden** is one of the best ways to ensure a steady supply of fresh, nutritious food during a long-term crisis. While it takes time to establish and maintain, a well-planned garden can provide you with a variety of fruits, vegetables, and herbs that will keep you nourished for months or even years.

Planning Your Garden

- **Location**: Choose a location that receives at least **6-8 hours of sunlight per day**. If you have limited space, you can also consider **container gardening** or building a **vertical garden** on a balcony or patio.
- **Soil**: Healthy soil is the foundation of any successful garden. If your soil is poor, you may need to **amend it with compost** or **organic fertilizers**. Alternatively, you can build **raised garden beds** and fill them with high-quality soil.
- **Crops**: Focus on growing **calorie-dense** and **nutritious crops** that are easy to grow and store. SEALs often rely on energy-dense foods like **potatoes, beans, and grains** in the field, and these should be staples in your survival garden. Other good options include:

- **Leafy Greens**: Spinach, kale, and lettuce grow quickly and provide essential vitamins and minerals.
- **Root Vegetables**: Carrots, beets, and sweet potatoes store well and can be grown in various climates.
- **Beans and Legumes**: These are an excellent source of protein and can be dried and stored for long periods.
- **Squash and Pumpkins**: These are calorie-dense and have a long shelf life when stored properly.
- **Herbs**: Grow medicinal and culinary herbs like **basil, oregano, thyme**, and **mint**. These can add flavor to meals and provide natural remedies for minor illnesses.

Extending Your Growing Season

- **Cold Frames and Greenhouses**: To extend your growing season into the colder months, consider building a **cold frame** or **greenhouse**. These structures trap heat and protect plants from frost, allowing you to grow food year-round.
- **Indoor Gardening**: If you live in an area with harsh winters, indoor gardening can provide a continuous source of fresh greens and herbs. Use **grow lights** and **hydroponic systems** to grow plants indoors without soil.

Raising Livestock and Poultry

For those with more space, raising **livestock** or **poultry** can provide a steady supply of protein in the form of **eggs**, **milk**, and **meat**. SEALs often rely on animal protein during extended missions, and having a small homestead setup can make your food supply much more resilient.

Chickens for Eggs and Meat

Chickens are one of the easiest and most productive animals to raise in a small-scale survival setup. They provide a continuous supply of **fresh eggs** and can also be raised for meat if necessary.

- **Coop Setup**: Build a secure chicken coop with adequate ventilation, roosting bars, and nesting boxes. Ensure that the coop is **predator-proof**, especially if you live in a rural area.
- **Feed and Foraging**: Chickens can forage for insects, grass, and weeds, which reduces your need for commercial feed. However, stock up on **chicken feed** to ensure they have enough food in lean times.

Goats for Milk and Meat

Goats are highly adaptable animals that can thrive in a variety of environments. They provide a consistent supply of **milk**, which can be used for drinking, making cheese, and even soap. Some breeds of goats can also be raised for meat.

- **Shelter**: Goats need a **small barn** or **shelter** to protect them from the elements. They also need secure fencing, as they are known for their ability to escape.
- **Feeding**: Goats are browsers, meaning they prefer to eat shrubs, weeds, and rough forage. This makes them easier to feed than other livestock, especially if you have access to wooded areas.

Preserving Food for Long-Term Storage

In addition to growing your own food, you'll need to learn how to **preserve** it to build up reserves that will last through the winter or times of scarcity. SEALs often rely on preserved rations in the field, and preserving your own food can ensure that nothing goes to waste during the growing season.

Canning

Canning is one of the best ways to preserve fruits, vegetables, and meats for long-term storage. When done properly, canned foods can last for **years** without refrigeration.

- **Water Bath Canning**: Use this method for **high-acid foods** like tomatoes, fruits, pickles,

and jams. The food is packed into jars and submerged in boiling water to create a seal.
- **Pressure Canning**: For **low-acid foods** like vegetables, meats, and soups, you'll need to use a **pressure canner**. This method uses high heat and pressure to kill bacteria and create a safe seal.

Dehydration

Dehydrating is a simple and effective way to preserve foods by removing their moisture content, which prevents bacteria and mold growth. Dried fruits, vegetables, and meats can last for months or even years when stored properly.

- **Electric Dehydrators**: These are the easiest and most reliable way to dehydrate food at home. Use them to dry fruits, vegetables, herbs, and even meat for making jerky.
- **Sun Drying**: If you don't have access to a dehydrator, you can use the **sun drying method** for fruits and herbs. Place thinly sliced fruits on a drying rack and leave them in the sun for several days, covering them with a mesh screen to protect from insects.

Fermentation

Fermenting foods is an ancient preservation method that also increases the nutritional value of certain foods.

Foods like **sauerkraut**, **kimchi**, and **yogurt** can be fermented and stored for months.

- **How to Ferment**: To ferment vegetables like cabbage, simply pack them tightly into a jar with salt, and allow them to ferment at room temperature for several days. Fermented foods provide beneficial **probiotics** that support gut health.

10.3 - Sustainable Energy Solutions: Powering Your Home Without the Grid

In a long-term bug-in scenario, **energy** becomes a critical resource. If the power grid is down for an extended period, you'll need alternative ways to power your home, keep your food fresh, and ensure basic comfort. SEALs are trained to generate their own power in the field, using **solar panels**, **battery packs**, and **generators**. You can implement similar systems at home to create a sustainable energy setup.

Solar Power

Solar power is one of the best long-term energy solutions, as it harnesses the sun's energy to generate electricity. Whether you're looking to power your entire home or just a few essential devices, solar power can provide a reliable source of energy during a crisis.

Solar Panels

- **Roof-Mounted Panels**: Installing **roof-mounted solar panels** is a long-term investment that can provide significant energy savings. These panels capture sunlight and convert it into electricity, which can be used to power your home or charge battery banks.
- **Portable Solar Panels**: For those who don't have the space or resources for a full solar array, **portable solar panels** are a great option. These panels are lightweight and can be set up anywhere with direct sunlight. Use them to charge **batteries**, **power banks**, or small devices like radios and lights.

Battery Storage

To maximize the efficiency of your solar power system, you'll need a way to store excess energy for use at night or on cloudy days. **Solar battery storage systems**, like the **Tesla Powerwall**, allow you to store solar energy for later use.

- **Deep Cycle Batteries**: If you don't have a full battery storage system, **deep cycle batteries** can be used to store smaller amounts of energy. These batteries are ideal for powering **appliances**, **lights**, or **medical devices** during a power outage.

Generators

A **backup generator** is another option for providing electricity when the grid is down. SEALs often rely on portable generators in the field, and having one at home ensures you can keep essential systems running in a crisis.

Gas Generators

Gasoline-powered generators are the most common type of backup generator. They are powerful enough to run **refrigerators, lights**, and other appliances, but they require a steady supply of fuel.

- **Fuel Storage**: Make sure to store enough **gasoline** to power your generator for several days or weeks. Gasoline should be stored in **approved containers** and kept in a **cool, well-ventilated area** away from the house.
- **Generator Placement**: Always place your generator **outdoors** in a well-ventilated area to prevent **carbon monoxide poisoning**. Use **extension cords** to run power into your home from the generator.

Propane and Solar Generators

If you want to avoid the need for gasoline, consider a **propane-powered generator** or a **solar generator**. Propane generators are cleaner and more efficient than gasoline models, and propane is easier to store for long periods.

- **Solar Generators**: A **solar generator** is powered by solar panels and stores energy in a **battery bank**. These generators are quiet and don't require fuel, making them a great option for those looking to reduce their reliance on fossil fuels.

10.4 - Waste Management and Sanitation Systems

In a long-term bug-in, waste management and sanitation can become major concerns, especially if **plumbing systems** fail. SEALs are trained to manage waste in remote environments, and it's essential to have systems in place to safely deal with human waste, garbage, and graywater during a crisis.

Composting Toilets

A **composting toilet** is an excellent solution for managing human waste when the plumbing system is down. These toilets use **sawdust** or **peat moss** to break down waste into compost, eliminating the need for water.

- **How It Works**: After each use, a small amount of sawdust or other organic material is added to the toilet. The waste is then composted over time,

reducing odors and creating a usable product for gardening or disposal.
- **Maintenance**: Composting toilets require regular maintenance, including emptying the compost chamber and ensuring proper ventilation. They are ideal for long-term bug-in scenarios where access to running water is limited.

Graywater Systems

Graywater refers to wastewater from **sinks, showers**, and **washing machines**. Setting up a **graywater system** allows you to reuse this water for irrigation or other non-potable purposes, reducing your overall water consumption.

- **Graywater Collection**: Install a **graywater diverter** in your plumbing system to direct water from sinks and showers into storage tanks or directly into your garden. Make sure to use **biodegradable soaps** and cleaners to prevent chemicals from contaminating your soil.
- **Irrigation**: Graywater can be used to water **non-edible plants** in your garden. Be sure to avoid using graywater on **edible crops** unless it has been properly filtered.

10.5 - Psychological Resilience: Adapting to Sustainable Living

While creating sustainable systems for water, food, and energy is critical for survival, it's equally important to cultivate the **mental resilience** needed to adapt to this new way of life. SEALs are taught to remain flexible and resourceful in any situation, and you'll need the same mindset to thrive during a long-term crisis.

Adapting to a New Routine

Sustainable living requires daily maintenance and commitment. Growing your own food, managing waste, and generating energy will become part of your regular routine. SEALs often work in harsh conditions for extended periods, and they learn to find satisfaction in the small victories, like securing clean water or making a successful harvest. Embrace the challenge of sustainable living and find **joy in the process** of building self-sufficiency.

Staying Motivated

Long-term crises can lead to **feelings of isolation** and **fatigue**, especially when you're focused on survival. It's important to stay **mentally engaged** and **motivated**. Set small goals for your sustainable living systems, whether it's expanding your garden, increasing your water storage, or improving your energy efficiency.

Celebrate each achievement, no matter how small, as a step toward greater resilience.

Building Community Support

One of the best ways to stay resilient during a crisis is to build a **support network** with your neighbors or local community. SEALs often rely on their teammates for support, and having a group of like-minded individuals to share resources, skills, and information can make a huge difference. Consider **bartering** with neighbors, sharing gardening tips, or pooling resources to build a stronger, more resilient community.

10.6 - Final Thoughts: Thriving Through Sustainability

Sustainable living isn't just about surviving a crisis—it's about thriving despite the challenges. By creating systems that provide for your **basic needs**, you can reduce your reliance on external resources and increase your chances of surviving a long-term emergency. Whether it's collecting rainwater, growing your own food, or generating solar power, these sustainable solutions will help you build a **self-sufficient home** that can withstand whatever challenges come your way.

Remember, survival is about more than just stockpiling supplies—it's about developing the skills, systems, and mindset necessary to adapt and thrive, even in the toughest circumstances. By following the principles of **resilience** and **self-reliance** that Navy SEALs embody, you'll be well-prepared for any crisis that arises.

Chapter 11: Mental and Emotional Resilience: Maintaining Psychological Strength in Long-Term Crises

In any long-term crisis, your physical survival depends not just on your supplies, preparedness, or defensive strategies but also on your **mental and emotional strength**. Navy SEALs are often placed in highly stressful, isolated, and dangerous situations, requiring them to develop extraordinary levels of **psychological resilience** to remain focused, calm, and functional even in the face of overwhelming adversity. This chapter will explore the mental and emotional challenges that arise during prolonged bug-in scenarios and how you can apply **SEAL mental toughness strategies** to stay grounded, focused, and resilient in times of crisis.

While physical preparedness is critical, psychological resilience is often the key factor in determining who thrives and who crumbles during long-term crises. Whether you're dealing with the stress of isolation, the fear of an uncertain future, or the emotional toll of maintaining your family's safety, understanding how to navigate these challenges is essential for your survival and well-being.

11.1 - Understanding Psychological Resilience: What Navy SEALs Teach Us

Psychological resilience is the ability to **adapt to stress, adversity, and challenges** while maintaining your mental health and emotional well-being. In the context of a long-term crisis, this means being able to handle the **constant uncertainty**, **threats**, and **hardships** that may arise without becoming overwhelmed or paralyzed by fear or stress.

Navy SEALs train extensively in **mental resilience**, learning to control their emotions, manage their fears, and remain focused on their mission no matter what challenges they face. They are taught that the body can withstand incredible physical strain, but the mind is what ultimately determines success or failure in survival situations. By adopting similar mental strategies, you can significantly improve your ability to cope with the stresses of a long-term crisis.

The Importance of Mental Toughness

Mental toughness isn't about suppressing emotions or ignoring fear. Rather, it's about developing the **inner strength** to face challenges head-on while staying composed and focused. SEALs are taught to see obstacles as opportunities to grow stronger, and they learn to embrace discomfort and uncertainty as part of the journey. In a bug-in scenario, this kind of mental toughness will help you overcome fear, manage stress, and make clear decisions when the stakes are high.

The Stress Response: Understanding Fight, Flight, and Freeze

In a crisis, your body naturally activates the **stress response**, also known as **fight, flight, or freeze**. This is your brain's way of preparing you to deal with danger, either by confronting it, escaping from it, or freezing to assess the situation. While this response is designed to keep you safe, it can sometimes lead to **panic, anxiety**, or **overwhelm** if not managed properly.

Navy SEALs learn to recognize their own stress responses and use specific techniques to keep themselves calm and focused. By understanding how your body reacts to stress and learning to manage that response, you can remain in control of your emotions and actions, even in the face of a crisis.

11.2 - Managing Stress: SEAL Techniques for Staying Calm and Focused

During a long-term crisis, stress can come from a variety of sources, including the fear of the unknown, physical exhaustion, isolation, and the constant need to make decisions under pressure. SEALs are trained to manage stress in some of the world's most extreme environments, and the techniques they use can help you stay calm, focused, and mentally resilient during a prolonged bug-in.

Controlled Breathing (Box Breathing)

One of the simplest and most effective techniques for managing stress is **controlled breathing**, also known as **box breathing**. SEALs use this technique to calm their nervous system and regain focus in high-stress situations.

- **How to Practice Box Breathing**:
    1. Inhale through your nose for **4 seconds**.
    2. Hold your breath for **4 seconds**.
    3. Exhale slowly through your mouth for **4 seconds**.
    4. Hold your breath again for **4 seconds**.
    5. Repeat the cycle until you feel your heart rate slow and your mind calm.

Box breathing works by engaging the **parasympathetic nervous system**, which helps bring your body out of the **fight-or-flight state** and into a calmer, more focused mode. Practicing this technique during moments of stress or anxiety can help you think more clearly and make better decisions.

Visualization and Mental Rehearsal

SEALs are also trained to use **visualization techniques** to mentally rehearse their missions and prepare for challenges before they occur. This technique can help you stay mentally sharp and ready to respond to a variety of scenarios during a bug-in.

- **How to Use Visualization**:
    - **Imagine success**: Close your eyes and visualize yourself successfully navigating a difficult situation, whether it's securing your home, rationing supplies, or protecting your family. Focus on the details—what you see, hear, and feel as you move through the scenario.
    - **Mentally rehearse your actions**: Visualize how you would respond to a specific challenge, such as a power outage, home invasion, or food shortage. Mentally walk through the steps you would take, including any problem-solving or decision-making that would be required.

By mentally preparing for different scenarios, you can reduce the impact of surprise or fear when those challenges arise, and you'll feel more confident in your ability to handle them.

Emotional Control: Responding, Not Reacting

In high-stress environments, it's easy to become **emotionally reactive**, allowing fear, anger, or frustration to dictate your actions. SEALs are taught to recognize their emotions and choose how they respond, rather than reacting impulsively.

- **The Pause**: When faced with a stressful or emotional situation, take a moment to **pause** before reacting. This allows your brain to shift from the **emotional** part (the amygdala) to the **rational** part (the prefrontal cortex), enabling you to make a clear, thoughtful decision.
- **Acknowledge Your Emotions**: Ignoring or suppressing emotions can lead to burnout or breakdowns. Instead, acknowledge your emotions, but don't let them control your actions. Remind yourself that fear, frustration, or anger are natural responses, but you have the power to choose how you act in spite of them.

11.3 - Overcoming Isolation and Loneliness: Maintaining Social Connections

One of the most challenging aspects of a long-term bug-in is the **isolation** that can come with it. Being cut off from social interactions, physical contact, or community can lead to **loneliness**, **depression**, and a decline in mental health. SEALs often operate in isolated environments for extended periods, sometimes going days or weeks without contact with others. To cope with this isolation, they use specific strategies to maintain their mental health and emotional well-being.

Staying Connected: Even When Physically Isolated

Even if you're physically isolated from others, maintaining some form of **social connection** is critical to your mental health. SEALs in the field often rely on **radio communication** with their team or commanding officers to stay connected and grounded. While your situation may not involve radios, there are ways to stay in touch with the outside world.

- **Use Technology**: If internet or phone service is still available, make it a point to check in with family, friends, or neighbors regularly. Whether it's a text message, phone call, or video chat, maintaining a connection with others can reduce feelings of isolation and remind you that you're not alone.
- **Two-Way Radios**: If the power grid is down or communication systems fail, use **two-way radios** (discussed in Chapter 8) to maintain contact with neighbors or nearby family members. Establishing a communication routine can help combat loneliness and provide emotional support during difficult times.

Creating a Routine and Structure

In isolation, days can start to blend together, leading to feelings of **disorientation** and **boredom**. SEALs maintain strict routines, even in isolated environments, to keep their minds sharp and maintain a sense of purpose.

- **Daily Schedule**: Create a **daily schedule** that includes activities such as meal preparation, exercise, maintenance tasks, and time for rest. Having a clear structure to your day provides a sense of control and purpose, which can help mitigate feelings of hopelessness or anxiety.
- **Set Goals**: Set small, achievable goals for each day or week, whether it's learning a new skill, working on a project, or improving your self-sufficiency systems. Completing tasks provides a sense of accomplishment and progress, which can be a powerful motivator during long-term isolation.

Mental and Emotional Engagement

During long-term crises, mental and emotional stagnation can be just as dangerous as physical inactivity. SEALs combat this by keeping their minds engaged with challenges, training, and learning. Keeping your mind active helps ward off feelings of depression and anxiety and allows you to stay sharp and alert.

- **Learn New Skills**: Take the opportunity to learn new survival skills or improve existing ones. Whether it's gardening, food preservation, or home repairs, learning something new can keep

your mind engaged and give you a sense of progress.
- **Stay Physically Active**: Physical exercise is not only critical for maintaining physical health but also for improving mental health. SEALs rely on regular exercise to manage stress and maintain mental clarity. Develop a simple **exercise routine** that you can do inside your home or yard, such as bodyweight exercises, yoga, or stretching.
- **Meditation and Reflection**: Use periods of quiet or downtime for **meditation** or **self-reflection**. Meditation helps calm the mind and reduce stress, while reflection can give you perspective on your situation and help you make clearer decisions moving forward.

11.4 - Handling Crisis Fatigue: Dealing with Prolonged Stress

One of the biggest challenges in a long-term crisis is **crisis fatigue**—the physical and mental exhaustion that comes from constantly being on high alert or under stress. SEALs are trained to push through exhaustion during missions, but they also understand the importance of **recovery** and **rest** in maintaining long-term resilience.

Recognizing the Signs of Crisis Fatigue

Crisis fatigue manifests as a range of physical and emotional symptoms, including:

- **Irritability and Frustration**: You may find yourself becoming easily irritated or frustrated by small things, such as noises, delays, or minor inconveniences.
- **Physical Exhaustion**: Prolonged stress can leave you feeling physically drained, even if you haven't been engaging in strenuous activity.
- **Difficulty Concentrating**: You may notice that it's harder to focus on tasks or make decisions, as your brain struggles to process information under continuous stress.
- **Emotional Numbness**: Some people experience emotional numbness or detachment, feeling like they are going through the motions without really connecting to what's happening around them.

Creating Time for Recovery

In high-pressure environments, SEALs use **micro-recovery periods** to manage their energy and ensure they don't burn out. Similarly, during a long-term bug-in, you need to build in time for **rest and recovery**, even if you feel the need to stay constantly busy or alert.

- **Sleep**: Sleep is one of the most critical components of recovery. During long-term crises,

sleep may be interrupted by stress or uncertainty, but it's essential to prioritize it as much as possible. Establish a **sleep routine** and stick to it, even if it means going to bed earlier or taking short naps during the day.
- **Rest Breaks**: Throughout the day, take **short breaks** to rest, relax, and reset your mind. These breaks can be as simple as sitting quietly for a few minutes, listening to music, or enjoying a cup of tea. The goal is to allow your brain and body time to recover from the constant stress of the crisis.

Setting Boundaries and Managing Expectations

In a crisis, especially one that lasts for an extended period, it's easy to feel overwhelmed by the constant pressure to solve problems, keep your household safe, and manage all aspects of survival. SEALs are trained to set boundaries, both mentally and physically, to prevent burnout.

- **Recognize Your Limits**: Be realistic about what you can accomplish in a day, and don't push yourself beyond your physical or mental limits. Understand that it's okay to prioritize rest and recovery over productivity.
- **Delegate Tasks**: If you're sharing the bug-in experience with family members or neighbors, **delegate tasks** to others. Sharing the workload

not only reduces your stress but also helps build a sense of community and cooperation.

11.5 - Cultivating Hope and Positivity in Dark Times

Maintaining **hope** and a sense of **purpose** during a long-term crisis is critical for mental resilience. SEALs are taught to stay focused on their mission, even in the face of overwhelming odds, and to find strength in the small victories that lead them closer to success. In a bug-in scenario, cultivating hope and staying positive can help you push through the darkest times.

The Power of a Positive Mindset

Studies have shown that people who maintain a **positive mindset** during difficult times are more likely to survive and thrive in crisis situations. This doesn't mean ignoring the reality of the situation or pretending that everything is fine, but rather finding ways to **focus on the positives** and **acknowledge progress**.

- **Gratitude Practice**: Each day, take a moment to reflect on something you're grateful for, whether it's having food to eat, the safety of your home, or the support of loved ones. **Gratitude practices** can help shift your focus away from fear and anxiety and remind you of the positive aspects of your life.

- **Celebrate Small Wins**: During a long-term crisis, even small victories should be celebrated. Whether it's successfully growing a new crop in your garden, securing clean water, or completing a home improvement project, these small wins add up and give you a sense of progress.

Building a Sense of Purpose

In survival situations, having a **sense of purpose** is one of the most powerful motivators. SEALs are driven by their mission, and having a clear purpose helps them push through adversity. You can cultivate the same sense of purpose during a bug-in.

- **Focus on What You Can Control**: There will be many things outside of your control during a long-term crisis, such as the state of the world, the duration of the emergency, or external threats. Instead of fixating on these uncontrollable factors, focus on what you **can control**—your actions, your mindset, and your immediate environment.
- **Take Care of Others**: Helping others can provide a powerful sense of purpose during a crisis. Whether it's caring for your family, checking on neighbors, or sharing resources, **acts of kindness** can boost your mood and reinforce the sense that you're contributing to something greater than yourself.

11.6 - Final Thoughts: The Strength of the Human Spirit

Mental and emotional resilience is often what separates those who survive from those who thrive during a long-term crisis. By adopting the mental toughness strategies used by Navy SEALs—managing stress, embracing challenges, staying connected, and cultivating hope—you can ensure that your mind remains strong and focused, no matter how difficult the situation becomes.

Remember, surviving a long-term bug-in isn't just about having the right supplies or physical defenses. It's about maintaining the **inner strength** to persevere, adapt, and find purpose in even the darkest of times. With the right mindset, you can turn even the most challenging crisis into an opportunity for growth, resilience, and success.

Chapter 12: Managing Medical Emergencies: Field Medicine and Healthcare Preparedness

In a long-term crisis, professional medical services may be delayed, inaccessible, or overwhelmed, leaving you to manage medical emergencies on your own. Whether you're dealing with injuries, infections, or chronic health conditions, it's crucial to be prepared with both **medical knowledge** and **essential supplies**. Navy SEALs are trained extensively in **field medicine** and trauma care to ensure that they can respond quickly and effectively to medical emergencies in remote and hostile environments. While your situation may not involve active combat, the ability to treat injuries, prevent infections, and manage healthcare in a long-term bug-in can be just as critical to survival.

In this chapter, we will explore how to manage a wide range of medical issues during a crisis, from basic first aid to more advanced emergency care. You'll learn how to build a comprehensive **medical kit**, how to treat common injuries and illnesses, and how to ensure that your household stays healthy and prepared for medical emergencies.

12.1 - Building a Comprehensive Medical Kit: Essential Supplies for Every Scenario

Having a well-stocked **medical kit** is the first step in preparing for any potential medical emergency during a long-term crisis. While you may already have a basic first aid kit, it's important to expand that kit to include **advanced medical supplies** that can handle more serious injuries or health issues. Navy SEALs carry advanced trauma kits in the field to treat everything from minor wounds to life-threatening injuries, and your bug-in medical kit should be equipped with the tools needed for a wide range of medical situations.

Basic First Aid Supplies

Your medical kit should begin with the **basic first aid supplies** that are essential for treating cuts, scrapes, burns, and other minor injuries. These are the items you will use most frequently, so be sure to stock up on enough to last through an extended crisis.

- **Adhesive Bandages**: Include a variety of **bandage sizes** to cover small cuts, blisters, or abrasions. Look for **waterproof** and **heavy-duty** options to ensure they stay in place during physical activity.
- **Gauze Pads and Rolls**: Sterile gauze is essential for covering larger wounds and absorbing blood. Stock up on both **4x4-inch**

pads and **rolls of gauze** for flexibility in wound dressing.
- **Medical Tape**: Keep a supply of **adhesive medical tape** to secure bandages and gauze. Look for hypoallergenic tape if anyone in your household has sensitive skin.
- **Antiseptic Wipes**: These are critical for **disinfecting wounds** to prevent infections. Include **alcohol wipes**, **hydrogen peroxide**, or **betadine** for cleaning cuts and scrapes before bandaging.
- **Antibiotic Ointment**: Stock tubes of **antibiotic ointment** (like **Neosporin**) to apply to wounds after cleaning. This helps prevent bacterial infections.
- **Elastic Bandages**: **ACE bandages** or similar elastic wraps are useful for supporting sprained joints, providing compression for swollen limbs, or holding splints in place.
- **Burn Cream or Gel**: Include **burn cream** or **aloe vera gel** to treat minor burns, which can happen during cooking or working with alternative heating methods.
- **Cold Packs**: **Instant cold packs** are useful for reducing swelling and pain in injuries such as sprains or bruises. These packs are activated by squeezing or shaking and provide immediate cold therapy.
- **Sterile Gloves**: Keep a box of **nitrile gloves** in your kit to protect both you and the patient during

wound treatment. Sterile gloves help reduce the risk of infection when handling open wounds.

Advanced First Aid and Trauma Supplies

For more serious medical emergencies, such as **heavy bleeding**, **broken bones**, or **severe burns**, you'll need advanced supplies that go beyond basic first aid. These items are essential for managing trauma in the absence of professional medical help.

- **Tourniquets**: A **tourniquet** is one of the most critical tools for stopping severe bleeding from an arm or leg. SEALs are trained to apply tourniquets in combat situations, and it's vital that you learn how to use one as well. Look for **CAT (Combat Application Tourniquets)**, which are widely used by military personnel and designed for ease of use.
- **Hemostatic Dressings**: In addition to tourniquets, **hemostatic dressings** like **QuikClot** or **Celox** are designed to stop severe bleeding by promoting rapid blood clotting. These dressings are applied directly to the wound and can save a life by controlling bleeding until further care is available.
- **SAM Splint**: A **SAM splint** is a lightweight, flexible splint that can be molded to support fractures or sprains. It's an essential tool for stabilizing broken bones and can be used for injuries to the arm, leg, wrist, or ankle.

- **Chest Seals**: A **chest seal** is used to treat **sucking chest wounds**, which occur when air enters the chest cavity through a puncture wound, often leading to a collapsed lung. Chest seals help prevent air from entering the wound and allow air to escape, stabilizing the injury until more advanced care can be administered.
- **Suture Kit or Steri-Strips**: For deep lacerations that may require stitching, include a **suture kit** or **Steri-Strips** (also called butterfly bandages) in your medical kit. These can be used to close wounds temporarily until professional medical help is available.

Medications and Preventive Supplies

Medications are a crucial part of any bug-in medical kit, as access to pharmacies may be limited or unavailable. Make sure your kit includes a variety of over-the-counter and prescription medications to handle pain, infections, and chronic conditions.

- **Pain Relievers**: Stock up on common pain relievers such as **ibuprofen**, **acetaminophen**, and **aspirin**. These medications can help reduce pain, inflammation, and fever.
- **Antihistamines**: Keep a supply of **antihistamines** like **Benadryl** or **Zyrtec** for allergic reactions, insect bites, or skin irritations.
- **Anti-Diarrheal Medications**: Dehydration caused by diarrhea can be dangerous in a crisis.

Include **anti-diarrheal medications** like **Imodium** to treat digestive issues.
- **Electrolyte Replenishment**: Include **oral rehydration salts** or **electrolyte powder** to help treat dehydration caused by illness, heat, or physical exertion.
- **Prescription Medications**: Ensure that any household members who rely on **prescription medications** have an adequate supply to last several months. Speak with your healthcare provider about obtaining extra doses for emergency situations.

12.2 - Managing Medical Emergencies: Essential Skills for Every Household

Having the right supplies is only part of the equation. To effectively manage medical emergencies during a long-term bug-in, you must also develop **practical medical skills**. Navy SEALs are extensively trained in **field medicine**, which allows them to stabilize injuries and treat patients until they can access more advanced medical care. While you don't need the same level of training, learning **basic emergency medical procedures** can significantly improve your ability to respond to injuries and illnesses when professional help isn't available.

How to Perform CPR

Cardiopulmonary Resuscitation (CPR) is a critical skill for saving the life of someone who has stopped breathing or whose heart has stopped beating. SEALs are trained to perform CPR under extreme conditions, and it's one of the most important life-saving skills you can learn.

- **Chest Compressions**: Place the heel of one hand on the center of the patient's chest, and place your other hand on top. Use your body weight to push down **2 inches deep** and perform **30 compressions** at a rate of **100-120 compressions per minute**.
- **Rescue Breaths**: After 30 compressions, give **two rescue breaths** by tilting the person's head back, pinching their nose, and breathing into their mouth until you see the chest rise. Continue alternating between compressions and breaths until professional help arrives or the person starts breathing again.

How to Control Severe Bleeding

Controlling **severe bleeding** is one of the most critical first aid skills you can develop. SEALs are taught to stop bleeding quickly and efficiently to prevent the patient from going into **shock** or losing too much blood.

- **Direct Pressure**: Apply **direct pressure** to the wound using a clean cloth or sterile gauze. Hold

firm pressure for several minutes without lifting the cloth to check the wound.
- **Tourniquet Use**: If the bleeding is from an arm or leg and cannot be controlled with direct pressure, apply a **tourniquet** about **2-3 inches above the wound**. Tighten it until the bleeding stops, and note the time it was applied.
- **Hemostatic Dressing**: If available, use a **hemostatic dressing** in conjunction with direct pressure. Apply the dressing to the wound and press firmly to stop the bleeding.

How to Treat Burns

Burns can occur in a variety of situations, from cooking accidents to fire-related emergencies. Knowing how to properly treat burns can prevent infections and reduce pain.

- **Cool the Burn**: Immediately cool the burn by running it under **cool (not cold) water** for at least **10 minutes**. Do not use ice, as this can damage the tissue further.
- **Cover the Burn**: After cooling, cover the burn with a **sterile, non-stick bandage** or **clean cloth**. Avoid using cotton balls or anything that could leave fibers in the wound.
- **Pain Management**: For mild burns, over-the-counter **pain relievers** like ibuprofen can help reduce pain and swelling. For more

serious burns, seek professional medical attention as soon as possible.

How to Splint a Broken Bone

Fractures and broken bones are common injuries during crises, especially when engaging in physical labor or defensive activities. SEALs are trained to **immobilize** broken bones to prevent further damage and reduce pain.

- **Immobilize the Area**: Use a **SAM splint**, rolled-up clothing, or any rigid material to immobilize the broken limb. Place the splint so that it extends **above and below the injured area**.
- **Secure the Splint**: Use **bandages, belts, or strips of cloth** to secure the splint in place. Be sure to tie the splint tightly enough to hold the bone in place, but not so tight that it cuts off circulation.
- **Elevate and Ice**: If possible, **elevate the injured limb** and apply an **ice pack** to reduce swelling.

How to Treat Shock

Shock occurs when the body's organs don't receive enough oxygen due to trauma, blood loss, or severe illness. It's a life-threatening condition that requires immediate attention.

- **Lay the Person Down**: Have the person **lie flat on their back**, and elevate their legs about **12 inches** to help improve circulation.
- **Keep Them Warm**: Use a **blanket** or jacket to keep the person warm. Shock often causes a sudden drop in body temperature.
- **Monitor Breathing**: Keep an eye on the person's breathing and **check their pulse regularly**. If they stop breathing, be ready to perform CPR.

12.3 - Treating Common Illnesses and Infections: Maintaining Health During a Crisis

During a long-term bug-in, illnesses and infections can become more frequent due to limited access to professional healthcare, stress, and compromised hygiene. SEALs are taught to manage a variety of illnesses and infections in the field, and understanding how to prevent and treat common medical issues is critical to maintaining your household's health during a crisis.

Dehydration

Dehydration is a serious condition that occurs when the body loses more fluids than it takes in. It can be caused by illness, physical exertion, or inadequate water intake, and it can be especially dangerous for children and the elderly.

- **Symptoms**: Look for signs of dehydration such as **dry mouth, dizziness, dark urine**, and **lethargy**. Severe dehydration can cause **confusion** and **rapid heart rate**.
- **Treatment**: Encourage the person to drink **small sips of water** or an **oral rehydration solution**. If they are unable to drink, apply **wet cloths** to their skin to help cool them down and hydrate them externally.

Infections and Sepsis

Untreated wounds, cuts, or burns can easily become infected, leading to more serious complications such as **sepsis**—a life-threatening condition that occurs when the body's immune system responds to an infection by attacking its own tissues.

- **Signs of Infection**: Look for **redness, swelling, heat**, and **pus** around a wound. If the person develops a **fever**, **chills**, or **rapid breathing**, the infection may be spreading.
- **Treatment**: Clean the wound thoroughly with **antiseptic** and apply **antibiotic ointment**. If the infection worsens, you may need to administer **oral antibiotics** if they are available.

Gastrointestinal Illnesses

During a crisis, especially when hygiene standards are difficult to maintain, gastrointestinal illnesses such as **food poisoning** or **diarrheal diseases** can become more common.

- **Symptoms**: Look for symptoms such as **vomiting, diarrhea, stomach cramps**, and **fever**. Severe cases of diarrhea can lead to dehydration if not treated promptly.
- **Treatment**: Ensure the person drinks plenty of **fluids** and administer **electrolyte replacements** to prevent dehydration. Use **anti-diarrheal medications** like Imodium to manage symptoms, but avoid giving medications that could interfere with the body's ability to eliminate toxins in cases of food poisoning.

Heat Exhaustion and Heatstroke

During a long-term crisis, especially in hot climates, **heat exhaustion** or **heatstroke** can become serious threats. These conditions occur when the body overheats and is unable to cool itself down properly.

- **Symptoms of Heat Exhaustion**: Look for **heavy sweating, weakness, nausea**, and **dizziness**. If untreated, heat exhaustion can progress to heatstroke.
- **Symptoms of Heatstroke**: **Heatstroke** is a medical emergency. Symptoms include

confusion, rapid heart rate, headache, and unconsciousness.
- **Treatment**: Move the person to a **cool, shaded area**, remove excess clothing, and apply **cold compresses** or **ice packs** to the neck, armpits, and groin. Have the person sip cool water. If heatstroke is suspected, seek medical help immediately.

12.4 - Preventive Care and Hygiene: Keeping Your Household Healthy

Preventing illness is always easier than treating it, especially during a crisis when medical resources may be limited. Maintaining good **hygiene** and practicing **preventive care** can help reduce the risk of infections, illnesses, and other health complications in a long-term bug-in scenario. SEALs know the importance of preventive care in the field, as even small lapses in hygiene can lead to dangerous illnesses that compromise their mission.

Personal Hygiene Practices

Maintaining personal hygiene is critical to preventing the spread of germs, bacteria, and illness, especially in close quarters. While water may be limited in a bug-in scenario, there are still ways to maintain hygiene.

- **Handwashing**: Regular **handwashing** is one of the most effective ways to prevent the spread of

illness. If water is limited, use **alcohol-based hand sanitizers** that contain at least **60% alcohol**.
- **Bathing**: Even in a crisis, it's important to maintain some form of **personal cleanliness**. Use **baby wipes** or **wet cloths** to wipe down your body, especially in areas where bacteria can grow, like underarms and groin.
- **Dental Hygiene**: Dental infections can become serious if left untreated, so be sure to maintain **oral hygiene** by brushing your teeth with **toothpaste** and a toothbrush, or using **baking soda** and water if toothpaste isn't available.

Sanitizing Your Environment

Keeping your living space clean is equally important for preventing illnesses. Regularly disinfecting surfaces, especially in areas like the kitchen and bathroom, can help prevent the spread of germs and bacteria.

- **Disinfecting Surfaces**: Use **disinfectant sprays**, **bleach solutions**, or **antibacterial wipes** to clean surfaces like countertops, door handles, and light switches. Regularly disinfect any surfaces that are frequently touched.
- **Waste Management**: Properly dispose of waste to prevent the buildup of bacteria and pests. Use **trash bags** to contain waste and, if necessary, create a **designated waste disposal area** outside your living space.

Boosting Immune Health

A strong **immune system** is your best defense against illness, so focus on maintaining your overall health through **nutrition**, **hydration**, and **exercise**.

- **Nutrition**: Eat a **balanced diet** rich in vitamins and minerals to support your immune system. If fresh food is limited, use **vitamin supplements** to ensure you're getting enough nutrients, especially **vitamin C**, **vitamin D**, and **zinc**.
- **Hydration**: Staying hydrated is critical for immune function. Drink plenty of water, and if possible, consume beverages that contain **electrolytes**.
- **Exercise**: Regular physical activity helps boost immune function and maintain overall health. Develop a simple **exercise routine** that you can do inside or around your home.

12.5 - Final Thoughts: Medical Preparedness as a Lifeline

Managing medical emergencies and maintaining health during a long-term crisis requires a combination of **preparation**, **knowledge**, and **resourcefulness**. Just as Navy SEALs are trained to treat injuries, prevent infections, and manage healthcare in the field, you too can develop the skills and systems needed to keep your household healthy and safe during a bug-in scenario.

By building a comprehensive **medical kit**, learning essential **first aid and trauma care skills**, and focusing on **preventive health practices**, you can ensure that you're prepared for any medical emergency that may arise. Remember, in a long-term crisis, your ability to manage healthcare effectively will be one of the most important factors in your overall survival and well-being.

Chapter 13: Long-Term Security and Protection: Evolving Your Defenses for Sustained Crisis Survival

In a long-term crisis, ensuring your home's security and your family's safety becomes an ongoing challenge. As resources become scarcer, the risk of opportunistic criminals, desperate individuals, and even organized groups targeting your home increases. Navy SEALs are experts in securing hostile environments, continuously adapting their defensive strategies to meet evolving threats. Similarly, during an extended bug-in, you must be prepared to **fortify your defenses**, **adapt your tactics**, and **protect your household** for the long haul.

This chapter will guide you through creating a dynamic security strategy that goes beyond basic home protection. We will explore how to reinforce your home's defenses, handle potential confrontations, and manage your visibility and resources to avoid drawing unwanted attention. As the crisis continues, threats may change in nature, making it crucial to balance **security** with **stealth** and **resilience**.

13.1 - Layered Defense Systems: Strengthening and Adapting Your Perimeter

In **Chapter 9**, we discussed the importance of establishing layers of defense around your home. In a long-term crisis, these layers must be **reinforced** and **adapted** to address new threats as the situation evolves. SEALs are trained to create **flexible defensive systems** that can be adjusted based on intelligence and changing environments, and you can apply this same principle to your bug-in scenario.

Reinforcing Your Perimeter: Evolving Threat Response

Your home's **perimeter** is the first line of defense against potential intruders. As the crisis drags on, external threats may become more frequent and more organized, requiring you to constantly evaluate and reinforce your perimeter security.

Barriers and Deterrents

Physical barriers play a crucial role in slowing down intruders and making it harder for them to breach your property. As the crisis extends, these barriers should be regularly inspected and improved.

- **Fence Reinforcement**: If you have a fence, consider adding additional layers of **security wire**, such as **barbed wire** or **razor wire**, along

the top. This will make it more difficult for intruders to climb over. Check regularly for signs of tampering or weak points.
- **Spiked Obstacles**: You can create low-tech obstacles such as **spiked wooden stakes** or **metal rebar stakes** embedded in the ground along the outer perimeter of your property. These can be concealed under foliage or natural debris to deter foot traffic.
- **Noise-Making Barriers**: Consider using gravel, pebbles, or **crunchy materials** around entry points and pathways near your perimeter. These create noise when stepped on, alerting you to the presence of intruders and making it difficult for them to approach stealthily.

Improving Visibility and Surveillance

Effective surveillance is essential for maintaining situational awareness and identifying threats before they reach your home. SEALs rely on **reconnaissance** and **constant monitoring** of their environment, and you should do the same.

- **CCTV Systems and Night Vision Cameras**: Upgrade your **security camera system** to include **night vision cameras** if you haven't already. Night is when most intruders attempt to breach defenses, and having reliable night vision will give you a tactical advantage.

- **Motion-Activated Lights**: Continue to expand your **motion-activated lighting** around your perimeter, especially in areas that are less frequently monitored. Ensure lights are placed out of reach to prevent tampering and that they cover key entry points such as **gates**, **driveways**, and **windows**.
- **Watch Points**: Set up **elevated watch points** or **lookout areas** where you can safely observe your property without exposing yourself to danger. These could be on rooftops, balconies, or in trees, giving you a clear view of the perimeter.

Managing Stealth: Avoiding Unwanted Attention

While fortifying your perimeter is essential, maintaining **stealth** is equally important. In a long-term crisis, drawing too much attention to your home can make you a target for those seeking resources or easy victims. SEALs often operate under strict rules of **noise and light discipline** to avoid detection, and you should apply similar principles.

Reducing Noise and Light Pollution

- **Quiet Home**: Avoid making unnecessary noise inside or outside your home, especially at night. Loud conversations, music, or construction sounds can signal to others that your home is occupied and possibly well-supplied.

- **Light Discipline**: Keep **curtains closed** at night and minimize the use of bright lights that could be seen from outside. Use **red or dim lighting** indoors if you need to move around at night, as this is harder to detect from a distance.
- **Generator Noise**: If you're using a **generator** to power your home, consider building a **soundproof enclosure** around it to reduce the noise. Unattended generator sounds can attract attention and may lead to a theft attempt.

Concealing Your Supplies and Activity

- **Camouflage Your Home**: In extreme situations, you may need to make your home appear **less appealing** to potential looters or attackers. Consider using **natural camouflage** by blending your home's exterior with surrounding vegetation or debris.
- **Resource Concealment**: If you have a garden, water collection system, or other visible supplies, do your best to **conceal them** from casual observation. Intruders who see abundant resources may be more likely to target your home.
- **Minimizing Foot Traffic**: Limit the number of times you leave your home to **retrieve supplies** or check on external systems like water collection. Frequent movement in and out of your home increases the chances that someone will notice your activity.

13.2 - Contingency Plans: Preparing for a Breach

Even with the best defensive strategies, there is always the possibility that your home's defenses could be breached by intruders. SEALs are trained to plan for **worst-case scenarios**, always having a **contingency plan** in place for escape, counterattack, or evacuation. You should adopt the same mindset, preparing for what to do if intruders manage to breach your defenses.

Safe Rooms and Retreat Areas

One of the most important elements of a contingency plan is having a **safe room** or **retreat area** where you can go in case of a breach. SEALs often use **fallback positions** during missions, and your home should have a designated safe area that is heavily fortified and equipped with essential supplies.

Choosing a Safe Room

Your safe room should be located in a part of the house that is **difficult to access** and easy to defend. Ideally, it should be an interior room without windows, such as a **bathroom**, **closet**, or **basement**. The door to the safe room should be **solid-core** and reinforced with a **deadbolt** or **security bar** to prevent intruders from breaking in.

Stocking Your Safe Room

- **Water and Food**: Ensure that your safe room is stocked with at least **24-48 hours' worth of water and non-perishable food**. This will give you enough time to stay secure until the intruders leave or help arrives.
- **Communications**: Keep a **radio**, **cell phone**, or **two-way radio** in the safe room so you can call for help or communicate with others in your household.
- **First Aid and Self-Defense Tools**: Include a **first aid kit** and **self-defense tools** (such as a **firearm**, **pepper spray**, or **tactical knife**) in your safe room in case you need to defend yourself.
- **Ventilation**: Ensure the room has some form of **ventilation**, especially if it's a sealed space, to avoid feeling trapped or suffocated.

Home Defense and Self-Defense Tactics

If intruders breach your home, it's critical that you are prepared to defend yourself and your family. Navy SEALs are experts in **close-quarters combat (CQC)**, a skill set that can be adapted for home defense. While firearms are often the most effective tool, knowing how to **position yourself**, **use the environment**, and **deploy non-lethal weapons** can make all the difference in surviving an intrusion.

Firearms for Home Defense

If you own a **firearm** and are trained in its use, it can be one of the most effective tools for defending your home. SEALs are taught to use firearms with precision and discipline, and you should apply the same principles.

- **Shotguns**: A **pump-action shotgun** is one of the best weapons for home defense, especially in close quarters. It provides stopping power and doesn't require pinpoint accuracy, making it ideal for dealing with intruders in confined spaces.
- **Handguns**: A **handgun** can be a useful secondary weapon, especially if you need something more portable. Keep it easily accessible but safely stored when not in use.
- **Safe Storage**: If you have firearms, ensure they are stored in a **locked safe** or **gun cabinet** but are still quickly accessible in an emergency.

Non-Lethal Self-Defense Options

Not everyone is comfortable using firearms, and in some cases, non-lethal self-defense options may be preferable.

- **Pepper Spray**: **Pepper spray** is a highly effective tool for incapacitating intruders without causing permanent harm. Keep cans of pepper spray in accessible locations around your home.
- **Stun Guns or Tasers**: **Stun guns** or **Tasers** provide a temporary incapacitation that can give

you time to escape or call for help. They are easy to use and don't require a high level of training.
- **Tactical Batons**: A **collapsible baton** can be used to strike an intruder and disable them temporarily. This tool is compact and can be easily carried with you inside the home.

Using the Environment to Your Advantage

- **Barricades**: Use furniture or other heavy objects to create **barricades** in hallways or doorways, slowing down intruders and giving you time to prepare a defense. SEALs often use obstacles to force enemies into a bottleneck, where they can be neutralized more effectively.
- **Choke Points**: Identify **choke points** in your home, such as narrow hallways or stairwells, where intruders will have limited mobility. These are ideal locations to set up defenses because you can limit the number of attackers who can approach at once.
- **Cover and Concealment**: In a confrontation, seek out **cover** (such as heavy furniture or walls) to protect yourself from gunfire or physical attacks. **Concealment** (such as hiding behind curtains or doors) can also be useful for avoiding detection, but remember that concealment doesn't provide physical protection.

13.3 - Handling Confrontations: SEAL Strategies for De-Escalation and Engagement

While defensive measures are critical, there may be times when you need to directly confront individuals who threaten your home. SEALs are trained in **negotiation**, **de-escalation**, and, when necessary, **engagement** with hostile forces. Knowing when and how to confront potential threats can make the difference between avoiding a dangerous situation and escalating into violence.

De-Escalating Threats: When Talking is the Best Strategy

In some cases, de-escalation is the safest and most effective way to resolve a confrontation. SEALs are often trained to negotiate or de-escalate tense situations before resorting to force, and this approach can be applied to home defense as well.

Body Language and Non-Verbal Cues

- **Maintain a Calm Posture**: Even if you feel threatened, try to maintain a calm and non-threatening posture. Standing too aggressively or making sudden movements can escalate the situation.
- **Use Open Hands**: When talking to potential intruders, keep your hands visible and open. This

signals that you're not holding a weapon and are willing to communicate.
- **Control Your Breathing**: Keep your breathing slow and controlled to help manage your stress levels and keep your voice steady.

Verbal Communication

- **Stay Calm and Assertive**: When speaking, remain calm but assertive. Avoid shouting or using threatening language, as this can escalate the situation.
- **Offer an Out**: Sometimes, intruders are looking for an easy target and may leave if they realize they're facing resistance. Offer them an opportunity to leave without conflict by stating something like, "We don't want trouble—please leave now."
- **Avoid Giving Away Information**: Don't reveal critical information such as the size of your household, the supplies you have, or your defensive capabilities. Keep your answers vague but firm.

When Engagement is Necessary: SEAL Combat Principles

In some cases, de-escalation won't be an option, and you may need to **engage** intruders to protect yourself and your family. SEALs are trained to engage with precision and control, minimizing risks to themselves and others.

Positioning and Tactical Advantage

- **Control the High Ground**: In any combat scenario, controlling the **high ground** gives you a tactical advantage. If possible, engage from elevated positions such as **stairs**, **balconies**, or **rooftops**.
- **Flanking and Angles**: If there are multiple intruders, try to **flank** them by moving to an angle that exposes their sides or backs. SEALs use flanking to neutralize threats from unexpected directions, limiting the intruder's ability to react.
- **Limit Their Mobility**: Use the environment to limit the intruder's mobility. Barricades, doorways, and narrow hallways can force them into predictable movement patterns, making it easier for you to engage from a position of strength.

Commit to Action

If you are forced to engage an intruder, do so with **complete commitment**. SEALs are trained to eliminate hesitation in combat, as hesitation can be fatal. Whether using a firearm or non-lethal weapon, act decisively and aim to neutralize the threat as quickly as possible.

13.4 - Adapting to Changing Threats: Long-Term Security Considerations

As a crisis stretches on, the types of threats you face may change. Early in the crisis, looters or opportunistic criminals may be your primary concern. However, over time, more organized groups, desperate individuals, or even governmental forces may pose different types of threats. SEALs are experts in **adapting their strategies** based on evolving intelligence, and you must be ready to do the same.

Threat Assessment and Intelligence Gathering

- **Monitor Your Surroundings**: Constantly monitor your surroundings for changes in the types of threats you may face. Use **scouting**, **surveillance**, and **local intelligence** (from neighbors, radios, or social media) to gather information about emerging dangers.
- **Reassess Your Defenses**: Periodically reassess your **defensive strategies** to ensure they remain effective. As new threats arise, you may need to upgrade your security measures, increase your surveillance, or strengthen your fallback positions.

Managing Resource Security

As the crisis continues, protecting your **supplies and resources** will become increasingly important. SEALs are trained to **secure critical assets** in hostile environments, and you should adopt similar strategies for your food, water, and other essentials.

- **Hidden Storage Areas**: Consider creating **hidden storage areas** within your home where you can conceal critical supplies. These can be false walls, floorboards, or hidden compartments that are not immediately visible to intruders.
- **Diversion Supplies**: Keep a small amount of **diversionary supplies** in an easily accessible location. If looters break in, they may take the easily visible supplies and leave without searching for your hidden reserves.
- **Guarding Critical Infrastructure**: Systems such as **water filtration**, **solar power**, or **gardens** are essential for long-term survival. Take extra care to secure and conceal these assets from potential attackers who may try to steal or destroy them.

13.5 - Final Thoughts: The Evolution of Home Security

As a crisis unfolds, the threats you face will evolve, requiring you to continually adapt your security strategies and defenses. Just as Navy SEALs remain flexible and responsive to changing environments, you too must be prepared to strengthen your perimeter, plan for worst-case scenarios, and engage with threats when necessary. By balancing **stealth**, **fortification**, and

tactical preparedness, you can create a secure home that withstands the challenges of long-term crisis survival.

Remember, protecting your home isn't just about building walls or collecting weapons—it's about staying **mentally sharp**, **tactically flexible**, and **ready for anything** that comes your way. With the right combination of defensive strategies, preparedness, and resilience, you can turn your home into an impenetrable stronghold capable of withstanding any threat.

Chapter 14: Building Community Networks: Leveraging Cooperation for Long-Term Survival

During a long-term crisis, isolation can increase your vulnerability. While it's essential to focus on fortifying your home and securing resources, working with others and building a **community network** can exponentially increase your chances of survival. Navy SEALs work in teams, leveraging each member's strengths and knowledge to accomplish complex missions. By applying similar principles to your crisis preparedness, you can pool resources, exchange vital information, and form alliances that will enhance your household's resilience.

In this chapter, we'll explore how to build trust with neighbors, form alliances, develop community roles, and manage conflicts that may arise. You'll learn how to balance cooperation with self-reliance, ensuring that your household remains secure while benefiting from the collective strength of a networked community.

14.1 - The Importance of Community in a Long-Term Crisis

Survival in a long-term crisis often depends on more than just individual preparedness. While Navy SEALs are trained to survive in harsh conditions alone, their missions are most successful when they work as a team. A well-coordinated community provides a support system that can reduce the strain on individual households, increase access to resources, and improve overall security.

Why Community Matters in a Crisis

There are several reasons why building a **community network** is vital for long-term survival:

- **Resource Sharing**: In a crisis, no single household can be fully self-sufficient indefinitely. **Bartering** and sharing resources—whether it's food, medical supplies, tools, or skills—help ensure that everyone in the network has access to essential goods.
- **Skill Exchange**: Each person in your community may have specialized skills that others lack, such as medical knowledge, mechanical skills, or gardening expertise. Sharing these skills helps cover gaps in individual preparedness.
- **Security Through Numbers**: A larger group provides **strength in numbers**, making it harder for opportunistic criminals or organized groups to target your community. Organized community patrols and watch systems can deter threats and keep everyone safer.
- **Emotional and Mental Support**: Extended crises take a toll on **mental health**. Being part of

a community offers emotional support, reducing the feelings of isolation, anxiety, and stress that often accompany long-term survival scenarios.

Balancing Community and Self-Reliance

While community offers many advantages, it's important to balance **cooperation** with **self-reliance**. Navy SEALs work as part of a team, but they are also trained to be completely self-sufficient in case they get separated from their unit. You should approach your community network with the same mindset: contribute to the group, but ensure that your household is still capable of surviving independently if necessary.

14.2 - Forming Alliances: Building Trust with Neighbors

The foundation of any successful community network is **trust**. In a crisis, people may be suspicious, scared, or unwilling to cooperate with others. Navy SEALs are trained to build relationships with local allies in hostile environments, often relying on **trust-building techniques** to form alliances. In your neighborhood or community, you'll need to establish trust before you can effectively work together.

Identifying Potential Allies

Start by identifying who in your neighborhood or local area could be a valuable ally. While it's tempting to trust friends or acquaintances, in a crisis, the most reliable allies are those who are **prepared** and **capable**. Look for people who have:

- **Practical skills**: Doctors, nurses, engineers, mechanics, and anyone with practical skills that could be useful during a long-term crisis.
- **Shared values**: Individuals or families who share similar values about preparedness, self-reliance, and security.
- **Access to resources**: Households with **tools**, **vehicles**, or **large properties** can contribute to the collective effort in ways that smaller, less-equipped homes cannot.

Building Trust in a Crisis

Trust is the foundation of any strong alliance, but it can be fragile, especially in high-stress situations. Here are some strategies for building trust with potential allies:

Start with Small Gestures

- **Share information**: One of the easiest ways to build trust is by sharing information. For example, if you learn of an emergency broadcast or a development in the crisis, share it with your neighbors. This shows that you are willing to help and keeps lines of communication open.
- **Exchange resources**: In the early stages of a crisis, offer to share small, non-critical resources, such as tools, water filters, or garden seeds.

Small gestures of cooperation build goodwill without jeopardizing your household's security.

Demonstrate Competence and Reliability

- **Be consistent**: In any crisis, reliability is key to building trust. Demonstrate that you can be counted on by showing up for meetings, sticking to agreed-upon schedules, and following through on promises.
- **Offer help**: If a neighbor needs assistance with something you're skilled in—such as fixing a broken generator, building a rainwater collection system, or setting up a garden—offer to help. This shows that you bring value to the community.

Develop Clear Communication Channels

Establishing **clear lines of communication** is critical for maintaining trust and cooperation within your network. Navy SEALs often rely on secure communication channels to stay in touch during missions, and your community network should have similarly reliable methods of communication.

- **Walkie-Talkies or Two-Way Radios**: Distribute **walkie-talkies** or **two-way radios** to trusted neighbors for easy communication during emergencies.

- **Regular Check-Ins**: Set up a schedule for regular check-ins, either by phone, radio, or in person. This keeps everyone informed and helps identify potential problems early.

Forming a Community Watch Group

One of the best ways to formalize your community network is by forming a **community watch group**. This group can focus on monitoring the neighborhood for potential threats, organizing resource-sharing efforts, and responding to emergencies.

- **Patrol Rotations**: Organize rotating patrols, where different households take turns monitoring the neighborhood. This ensures constant vigilance without overburdening any one family.
- **Emergency Signals**: Develop a system of **emergency signals** (such as flashing lights, radio codes, or whistles) to alert the group to immediate threats.

14.3 - Organizing Roles and Responsibilities: A Team Approach to Survival

Once you've established a basic level of trust with your community, the next step is to **organize roles and responsibilities** within the group. SEAL teams are highly organized, with each member taking on specific roles based on their skills and strengths. Similarly, your community network should divide tasks and responsibilities to ensure that all aspects of survival are covered.

Assigning Roles Based on Skills

Each member of the community brings different strengths to the table, so it's essential to **assign roles** based on expertise. Here are some common roles that can be assigned within a community network:

- **Security and Defense**: Individuals with experience in law enforcement, military, or firearms training can take the lead in organizing security, including patrols, defensive strategies, and weapons training.
- **Medical Care**: Those with medical knowledge, such as doctors, nurses, or paramedics, should be responsible for providing first aid, treating illnesses, and managing any medical emergencies.
- **Food and Water Management**: Gardeners, hunters, and those skilled in **food preservation** should oversee food production, storage, and rationing. Water purification experts or engineers can manage the community's water supply.
- **Communications Coordinator**: Someone with strong organizational skills should take charge of **communication channels**, ensuring that information is distributed efficiently, and that the group stays in contact.

- **Logistics and Resource Management**: A logistics coordinator can manage the group's **supply chain**, including inventorying stockpiles, organizing bartering, and coordinating the acquisition of necessary supplies.

Rotating Responsibilities to Prevent Burnout

Long-term crises can lead to **burnout**, especially if certain individuals are overburdened with too many responsibilities. Navy SEALs rotate leadership roles and responsibilities in high-stress missions to avoid fatigue, and your community should do the same.

- **Patrol Rotations**: As mentioned earlier, rotate patrol duties among different households to ensure that no one is overburdened.
- **Task Sharing**: If a role becomes too demanding (such as managing food production or medical care), consider **dividing** that role between two or more individuals.

Planning for the Long-Term: Sustainable Roles

The longer a crisis lasts, the more important it is to ensure that roles within the community are **sustainable**. People may get sick, injured, or simply exhausted, so it's essential to have **backup plans** in place for critical roles.

- **Cross-Training**: SEALs are trained in multiple disciplines, so that if one team member is

incapacitated, others can step in. Apply this principle by cross-training community members in basic medical care, food production, or security.
- **Apprenticeships**: Encourage those with specialized skills to take on **apprentices**, so that knowledge can be passed down and more people can contribute to essential tasks.

14.4 - Sharing Resources and Bartering: Managing Community Supplies

Resource management is one of the most critical aspects of long-term survival. In a prolonged crisis, external supplies may become scarce, forcing communities to rely on **resource-sharing** and **bartering** to meet their needs. SEALs often operate in environments with limited resources, making efficient supply management essential for mission success. By pooling and sharing resources within your community network, you can ensure that everyone has access to essential goods without depleting your own stockpiles.

Establishing a Community Resource Pool

Pooling resources—whether it's food, water, medical supplies, or tools—allows the community to make the most of what's available. However, this requires a high level of trust and organization.

What to Pool

- **Non-Perishable Foods**: Encourage households to contribute non-perishable foods such as canned goods, dried grains, and freeze-dried meals. These items can be rationed over time.
- **Medical Supplies**: Create a **community medical kit**, where individuals can contribute extra medical supplies like bandages, antiseptics, and over-the-counter medications.
- **Tools and Equipment**: Pool **tools** such as hammers, saws, and axes for building or repairing critical infrastructure. Larger equipment like **generators** or **water filtration systems** can also be shared.
- **Seeds and Gardening Supplies**: Share gardening tools, seeds, and knowledge to help more households start their own food production systems.

Rationing and Distribution

Once resources are pooled, the group will need to establish a **fair system** for distributing them. SEAL teams rely on precise logistics and rationing to ensure that everyone gets what they need, and your community should do the same.

- **Rationing Plan**: Develop a **rationing plan** that allocates food, water, and medical supplies based on the number of people in each household. Make sure that the plan accounts for

special needs, such as children, the elderly, or those with medical conditions.
- **Emergency Reserves**: Set aside **emergency reserves** of food, water, and medical supplies that can be accessed only in extreme situations. These reserves should be closely guarded and used sparingly.

Bartering Within the Community

In a crisis, **bartering** often replaces traditional forms of commerce. SEALs are trained to barter and trade for critical supplies when deployed in foreign territories, and this skill becomes invaluable when external resources are limited.

- **What to Barter**: Items that are in short supply—such as **fuel, ammunition, medical supplies**, or **food**—can be used for bartering. Skills and labor (such as **medical treatment**, **construction**, or **repair work**) are also valuable barter commodities.
- **Fair Trades**: Establish a **barter system** within the community that ensures fair trades. This could be as simple as a barter market where people exchange goods and services, or a more organized system with a designated barter coordinator.

Managing Conflicts Over Resources

One of the biggest challenges in a resource-sharing community is managing **conflicts** that arise when supplies run low. SEAL teams are trained to handle interpersonal conflicts diplomatically, and you'll need to apply similar conflict-resolution skills within your community.

- **Transparency**: Ensure that all decisions about resource allocation are made **transparently** and **fairly**. Hidden stockpiles or preferential treatment can lead to resentment and distrust.
- **Mediation**: If conflicts arise over resources, appoint a neutral mediator to help resolve the issue. This could be someone who is respected within the community or an individual with experience in conflict resolution.
- **Restorative Agreements**: If someone violates the community's rules regarding resource sharing (such as hoarding supplies or stealing), focus on **restorative justice**—repairing the harm done—rather than punitive measures. This helps maintain trust and unity within the group.

14.5 - Handling Community Challenges: Conflict Resolution and Leadership

Even in the most cooperative communities, **conflicts** are bound to arise. SEAL teams operate in high-stress environments, where interpersonal tensions can flare, but they rely on strong leadership and conflict resolution skills to maintain unit cohesion. In your community network, handling conflicts quickly and diplomatically is essential for keeping the group united and focused on survival.

The Role of Leadership

Effective leadership is critical in any community, especially during a crisis. While SEAL teams operate with a clear chain of command, your community may need to adopt a more flexible leadership model.

Choosing a Leader

The community should select a **leader** or **leadership council** to guide decision-making, mediate disputes, and organize community efforts. Leadership should be based on:

- **Experience**: Choose leaders with practical experience in crisis management, security, or logistics.
- **Respect**: Leaders should be respected by the community and capable of mediating conflicts without favoritism.
- **Adaptability**: Crisis situations are constantly evolving, so leaders must be able to adapt quickly and make decisions under pressure.

Rotating Leadership Roles

To prevent burnout and ensure that multiple people are prepared to lead, consider **rotating leadership roles** every few weeks or months. This allows different individuals to contribute their skills and ensures that the group remains cohesive if a leader is incapacitated.

Managing Community Conflicts

In any group, especially under the stress of a long-term crisis, conflicts will arise. SEAL teams are trained to resolve conflicts quickly to maintain mission focus, and your community should take a similarly pragmatic approach to conflict resolution.

Common Sources of Conflict

- **Resource Distribution**: Disagreements over how food, water, or medical supplies are distributed can lead to resentment and division.
- **Leadership Disputes**: Conflicts can arise if members of the community feel that leadership is acting unfairly or making poor decisions.
- **Personality Clashes**: Personal disagreements, misunderstandings, or cultural differences can create tension within the group.

Conflict Resolution Strategies

- **Mediation**: Use neutral mediators to help resolve conflicts between individuals or households. The

mediator should listen to both sides and help the parties reach a mutually acceptable solution.
- **Community Meetings**: Hold regular **community meetings** where grievances can be aired and resolved in a structured, respectful way. Allowing people to express their concerns prevents issues from festering and escalating.
- **Problem-Solving Mindset**: Encourage a **problem-solving mindset** in the group. Instead of focusing on blame or punishment, focus on finding solutions that benefit the entire community.

14.6 - Final Thoughts: The Strength of a United Community

In a long-term crisis, **community** becomes one of the most valuable assets you can cultivate. By building strong alliances, organizing roles, sharing resources, and managing conflicts effectively, you can create a network that enhances everyone's chances of survival. Navy SEALs understand that no mission is accomplished alone—teamwork, trust, and cooperation are critical to success. In your own crisis scenario, working with others will give you the strength, resilience, and support needed to thrive in even the most challenging situations.

By applying SEAL principles of leadership, teamwork, and resource management to your community network, you can build a **self-sustaining, cohesive group** that stands strong against any threat. Remember, in the end, survival is not just about individual preparedness—it's about **collective resilience**.

Chapter 15: Psychological Warfare: Defending Against Manipulation, Coercion, and Intimidation

Surviving a long-term crisis is not just about physical preparedness. There's another, often overlooked, aspect of survival that can be just as critical: **psychological warfare**. In times of crisis, desperation can lead people to use manipulation, coercion, and intimidation to get what they want. Whether it's controlling access to resources, spreading fear, or using deceit to weaken others, psychological tactics are a very real threat in survival situations.

Navy SEALs are trained to resist psychological manipulation, both in the field and when under enemy control. Their mental toughness, ability to identify manipulation tactics, and strategies for defending against coercion can be applied to civilian life during a crisis. This chapter will cover how to **recognize psychological threats**, **build mental defenses**, and **employ counter-tactics** to ensure you and your family remain resilient against emotional and psychological manipulation.

15.1 - Understanding Psychological Warfare in a Crisis

Psychological warfare refers to the use of **intimidation, fear, manipulation**, and **coercion** to influence people's behavior and decisions. In a long-term crisis, resources are scarce, tensions run high, and people may resort to these tactics to get what they want. Whether it's someone trying to intimidate you into giving up supplies, spreading false information to control a community, or manipulating alliances to their advantage, understanding the various forms of psychological warfare is crucial for survival.

Why Psychological Warfare Happens in Crises

Psychological warfare becomes a prominent tactic in crises for several reasons:

- **Scarcity of Resources**: When resources like food, water, and medicine become scarce, people may resort to manipulation and intimidation to secure their share or take from others.
- **Desperation and Fear**: Fear drives much of human behavior in a crisis. Desperate people often use fear tactics—threats, lies, and intimidation—to gain control or protect themselves.
- **Breakdown of Social Order**: As societal structures and laws break down, the usual

checks and balances that prevent people from engaging in aggressive or manipulative behavior may disappear, leading to more unscrupulous tactics.

Forms of Psychological Warfare

Psychological warfare in a crisis can take many forms, including:

- **Intimidation**: Using **threats of violence** or coercion to force people into compliance. This could involve direct threats or simply creating an atmosphere of fear.
- **Manipulation**: Deceptive tactics designed to make people act against their interests, such as spreading misinformation, exploiting emotions, or creating false alliances.
- **Divide and Conquer**: Creating division within a group or community to weaken its cohesion. This could involve pitting people against one another or sowing distrust within the group.
- **Emotional Manipulation**: Using guilt, sympathy, or shame to control others, such as pressuring someone to give up supplies by appealing to their sense of empathy.
- **Disinformation**: Spreading **false information** to cause confusion, panic, or doubt, making people question their decisions or follow misleading advice.

15.2 - Recognizing Psychological Manipulation and Intimidation Tactics

One of the most important steps in defending against psychological warfare is being able to **recognize the signs** of manipulation and intimidation. Navy SEALs are trained to be hyper-aware of their environment and the intentions of those around them. By developing a similar level of awareness, you can better protect yourself and your family from those who may try to exploit or manipulate you during a crisis.

Spotting Intimidation Tactics

Intimidation is often one of the first forms of psychological warfare you'll encounter in a crisis. The goal of intimidation is to instill fear, making you more likely to comply with the intimidator's demands.

- **Physical Posturing**: Intimidators often use **physical posturing**—standing too close, using aggressive body language, or making direct eye contact—to assert dominance. They may also use their size or numbers to make you feel outnumbered or overwhelmed.
- **Verbal Threats**: Direct or implied **verbal threats** are a common intimidation tactic. The individual may hint at violence, consequences, or retribution if you don't comply with their demands.

- **Fear of Loss**: Intimidators may attempt to exploit your fear of losing something important—such as supplies, security, or loved ones. For example, they may threaten to take food or harm family members if you don't comply.
- **Creating Urgency**: By creating a sense of **urgency** or a "now or never" scenario, the intimidator pressures you to make decisions without thinking them through.

Recognizing Manipulation Tactics

Manipulation is more subtle than intimidation and often relies on **deception, emotional exploitation**, or **distorted information** to influence your behavior. Manipulators may try to gain your trust before exploiting you.

- **Appealing to Emotion**: Emotional manipulation often involves appealing to your sense of **guilt, sympathy, or compassion**. For example, someone might feign vulnerability or desperation to manipulate you into sharing your resources.
- **Flattery and Trust-Building**: Manipulators often use **flattery** or false friendship to build rapport. They may offer small favors, gifts, or compliments to gain your trust before exploiting it.
- **Spreading Disinformation**: Manipulators may spread **false information** to mislead you into making poor decisions. For example, they might

claim there's a dangerous threat nearby to cause panic or manipulate you into following their lead.
- **Creating False Choices**: Manipulators may present you with a **false dilemma**, where it seems like you only have two options—both of which benefit them. For example, they might say, "Either you give me half your food, or I'll take all of it."

Understanding "Divide and Conquer" Tactics

The "divide and conquer" strategy is used to **weaken a group** by creating internal conflict or distrust. In a crisis, this tactic can be employed to sow division within a community or household, making it easier for an outsider to control or exploit the group.

- **Sowing Distrust**: Manipulators may spread **rumors** or **lies** to create suspicion between members of a group. For example, they might suggest that someone is hoarding supplies or planning to betray the group.
- **Exploiting Personal Conflicts**: If there are existing tensions or disagreements within your group, manipulators may try to **amplify these conflicts** to break the group apart. They might play both sides, pretending to sympathize with each faction.
- **Targeting Leaders**: Manipulators may try to **discredit** or **undermine leaders** by questioning their decisions, spreading false information, or

creating situations that make the leader appear weak or incompetent.

15.3 - Building Mental Defenses: Resisting Psychological Manipulation and Coercion

Once you can recognize psychological warfare tactics, the next step is to **build mental defenses** that will allow you to resist these manipulations. Navy SEALs undergo extensive mental toughness training to resist psychological pressure, maintain focus, and make clear decisions under stress. You can apply similar techniques to protect yourself from manipulation, intimidation, and coercion during a crisis.

Mental Toughness and Confidence

One of the best defenses against psychological warfare is **mental toughness**—the ability to stay calm, focused, and confident in the face of adversity. SEALs build mental toughness through rigorous training, but there are practical ways you can develop this skill at home.

Breathing Techniques and Mindfulness

When faced with intimidation or manipulation, it's easy to let fear or anger take over, leading to rash decisions. **Controlled breathing techniques** can help you stay calm and prevent emotional reactions.

- **Box Breathing**: This technique, used by SEALs, involves inhaling deeply for four counts, holding your breath for four counts, exhaling for four counts, and holding again for four counts. This rhythmic breathing calms the nervous system and helps you maintain focus.
- **Mindfulness Practices**: Practicing mindfulness allows you to observe your emotions without reacting to them immediately. This can be especially helpful when dealing with manipulative individuals, as it helps you stay aware of the situation without becoming overwhelmed by emotion.

Positive Visualization

SEALs use **positive visualization** to mentally rehearse successful outcomes before they happen. In the context of psychological warfare, you can use this technique to **imagine yourself resisting manipulation** or intimidation. Visualize staying calm, confident, and in control, even when faced with pressure. This mental rehearsal makes you more likely to act decisively in the moment.

Critical Thinking and Skepticism

Psychological manipulation often relies on **deception** or **false information**, so developing strong **critical thinking** skills is essential. SEALs are taught to question everything in hostile environments, ensuring they're never misled by misinformation.

Question Everything

When someone presents you with information—whether it's a warning about a threat, a request for supplies, or an emotional appeal—pause and ask yourself:

- **What is their motive?** What do they stand to gain by telling you this? Are they trying to manipulate your emotions or judgment?
- **Is this information verifiable?** Can you confirm what they're saying through other sources or evidence, or are they relying on your ignorance or fear?
- **What are the alternatives?** Are they presenting a false choice or trying to rush you into a decision without considering all your options?

The Power of "No"

Don't be afraid to say **no** when someone tries to manipulate you. Manipulators often rely on people's discomfort with confrontation, assuming you'll go along with their demands to avoid conflict. Practice saying no calmly and assertively, without providing long explanations or apologies. Remember, you are under no obligation to give away your resources or make decisions that compromise your safety.

15.4 - Countering Psychological Warfare: SEAL Strategies for Turning the Tables

In some situations, defending against psychological warfare isn't enough—you may need to **counter** these tactics to protect yourself and your household. SEALs are trained not only to resist manipulation but also to use psychological tactics of their own when necessary. By understanding how to turn the tables on those trying to manipulate or intimidate you, you can gain the upper hand.

Control the Narrative: Shaping Perception

One of the most powerful psychological tactics is **controlling the narrative**—shaping how others perceive you, your household, and your community. If you can control how others see you, you can reduce the likelihood of becoming a target and increase your influence within the community.

Appear Strong, Even When You're Not

SEALs are taught to **project strength and confidence**, even in difficult situations. In a crisis, appearing weak or vulnerable can make you a target for manipulation or exploitation.

- **Body Language**: Stand tall, make eye contact, and use confident body language, even if you

feel uncertain. This signals that you are in control and not easily intimidated.
- **Control Information**: Be selective about what information you share with others. Don't reveal your full resources, capabilities, or plans. This keeps potential manipulators guessing and prevents them from using your vulnerabilities against you.

Misinformation as a Defense

In some situations, **misinformation** can be used as a defensive tactic. By controlling what others know about your household, you can protect your resources and safety.

- **Downplay Your Supplies**: If someone suspects that you have more supplies than they do, downplay the amount you have. For example, if someone asks how much food you have, you can say, "We're running low too," to discourage further probing.
- **False Vulnerability**: In certain cases, appearing weaker than you actually are can deter more aggressive forms of manipulation. For instance, if a group is looking for targets, you might suggest that your household is struggling just like everyone else, making it less appealing to exploit.

Deflecting and Reversing Manipulation

SEALs are taught to **deflect** and **reverse manipulation** tactics, using the enemy's own strategies against them. In a survival scenario, these techniques can help you avoid being cornered by psychological warfare.

Turn the Focus Back on the Manipulator

When someone tries to manipulate you with emotional appeals or deceptive tactics, reverse the focus back on them:

- **Ask Questions**: Instead of responding directly to their request or statement, ask probing questions that make them explain themselves. For example, if someone tries to guilt you into sharing supplies, ask, "Why didn't you prepare ahead of time?" or "What did you do with your resources?"
- **Challenge Their Logic**: Manipulators often rely on flawed logic or emotional reasoning. Politely challenge their assumptions by pointing out contradictions or gaps in their reasoning. This can make them hesitate and rethink their strategy.

Offer an Unattractive Alternative

If someone tries to pressure you into making a choice that benefits them, present an **unattractive alternative**.

This shifts the power back to you and forces the manipulator to reconsider their demands.

For example, if someone demands half of your food supplies, you can say, "I can't give you half, but I'll give you a little now if you agree to help with security patrols in the future." This puts the burden of continued cooperation on them and discourages further manipulation.

15.5 - Protecting Your Community from Psychological Warfare

While protecting yourself from psychological warfare is essential, it's also important to safeguard your **community**. Manipulators may attempt to exploit divisions within your group, create mistrust, or sow confusion to weaken your collective strength. SEALs know that a divided team is a vulnerable team, and maintaining unity is critical to mission success.

Establish Clear Communication Channels

One of the most effective ways to prevent psychological manipulation within your community is to establish **clear, open communication channels**. SEAL teams rely on regular communication to ensure that everyone has the same information and can act cohesively. In your community, use similar tactics to prevent disinformation from spreading.

- **Regular Meetings**: Hold regular community meetings where important information is shared openly. This prevents rumors from taking hold and ensures that everyone is on the same page.
- **Fact-Checking**: Encourage a culture of **fact-checking**. Before acting on any information, verify its accuracy through multiple sources. This helps prevent panic and confusion from taking over.

Foster Unity and Trust

Divide-and-conquer tactics rely on exploiting internal divisions, so fostering **unity and trust** within your community is essential. SEAL teams build trust through shared experiences, mutual respect, and transparency, and your community can do the same.

- **Address Conflicts Early**: If conflicts arise within the group, address them **early and openly**. Avoid letting grievances fester, as unresolved conflicts can be easily exploited by manipulators.
- **Build Shared Goals**: Create **shared goals** that everyone in the community can work toward, such as improving security, increasing food production, or organizing patrols. These goals reinforce a sense of unity and purpose.

Neutralizing External Threats

If an outsider attempts to manipulate or coerce your community, it's important to **neutralize the threat** without allowing it to destabilize the group.

- **Confront the Manipulator**: If someone is trying to divide your group or spread false information, confront them directly but calmly. Publicly challenge their statements and demand evidence for their claims. This can expose their tactics and weaken their influence.
- **Exile as a Last Resort**: In extreme cases, if an individual is consistently undermining the group's safety or cohesion, you may need to **exile** them from the community. However, this should be a last resort, as it can have long-term consequences for group morale and trust.

15.6 - Final Thoughts: Fortifying Your Mind for Survival

Psychological warfare is a powerful tool, especially in times of crisis when emotions run high and people are more vulnerable to manipulation and fear. Just as Navy SEALs prepare their minds for the challenges of combat and high-stakes operations, you must prepare your mind to resist the emotional and psychological pressures that come with long-term survival scenarios.

By building mental toughness, recognizing manipulation tactics, and using counter-strategies, you can defend yourself, your family, and your community from those who seek to exploit or intimidate you. Remember, survival isn't just about physical strength—it's about maintaining control over your mind, your emotions, and your decisions in the face of uncertainty and adversity.

With the right mental defenses, you can turn psychological warfare into a weapon that strengthens your resolve and ensures your long-term resilience.

Chapter 16: Long-Term Sustainability: Adapting Your Home and Lifestyle for Indefinite Survival

In the early stages of a crisis, short-term solutions can keep you and your household safe and secure. But as time drags on, you may find that the initial supplies and strategies that worked for the first weeks or months of a bug-in are no longer sufficient. This is where **long-term sustainability** becomes critical. Much like Navy SEALs who operate in prolonged missions in hostile environments, you must learn how to adapt your home, lifestyle, and mindset to survive for as long as it takes.

This chapter will guide you through strategies for **sustainable living**, **resource management**, and **psychological endurance** that will ensure you not only survive but thrive in an extended crisis. From growing your own food to cultivating mental resilience, we'll cover the steps you need to take to make your household self-sufficient for the long haul.

16.1 - Transitioning from Short-Term to Long-Term Survival

In the initial phase of a crisis, the focus is on securing immediate needs like **food, water, shelter**, and

security. Most people rely on stockpiles, emergency kits, and short-term measures to get through the first few weeks. But as the crisis extends beyond the anticipated timeline, these resources will begin to dwindle, and it becomes clear that more permanent solutions are necessary.

Recognizing When to Transition to Long-Term Planning

A key aspect of long-term survival is recognizing the **signs** that it's time to transition from short-term crisis management to long-term sustainability. SEALs are trained to constantly reassess their environment and resources, ensuring they are prepared for the next phase of any mission. Similarly, you must remain vigilant and flexible, ready to adapt your strategy as the crisis evolves.

Signs It's Time to Shift Your Focus:

- **Depleting Resources**: If you notice your stored food, water, or fuel supplies running low, it's time to start looking for renewable sources and ways to conserve what you have.
- **Increased Isolation**: As weeks turn into months, access to external resources—whether through supply chains, community aid, or external help—may become increasingly scarce. The longer the isolation, the more critical it is to build self-reliance.

- **Escalating Security Threats**: If security conditions worsen over time, you may need to fortify your defenses further and take additional steps to protect your household.
- **Emotional and Mental Fatigue**: Long-term crises take a toll on mental health. If you or your household members are showing signs of **burnout, stress**, or **depression**, it's time to implement strategies for psychological resilience and well-being.

Mindset Shift: From Survival to Thriving

In a long-term crisis, the mindset shift from simply **surviving** to **thriving** is essential. Navy SEALs understand that endurance isn't just about hanging on—it's about adapting, overcoming challenges, and finding ways to succeed under pressure. By shifting your focus from short-term emergency measures to long-term sustainability, you can reduce stress, conserve resources, and create a lifestyle that supports ongoing resilience.

16.2 - Sustainable Food Production: Growing and Preserving Your Own Food

In a long-term crisis, one of the most pressing challenges is ensuring a continuous supply of food. While stockpiles can provide sustenance for weeks or even months, they will eventually run out. This is where **sustainable food production** comes into play. By growing your own food and learning preservation techniques, you can create a food system that will sustain your household indefinitely.

Starting a Survival Garden

One of the best ways to ensure long-term food security is by establishing a **survival garden**. SEALs operating in remote environments often rely on foraging and limited agricultural practices to sustain themselves, and you can apply similar principles to your home environment.

Choosing What to Grow

When starting a survival garden, it's important to choose crops that are **nutrient-dense**, **calorie-efficient**, and relatively easy to grow in your climate. Focus on plants that are resilient, store well, and can provide a steady food supply throughout the year.

- **Root Vegetables**: Potatoes, sweet potatoes, carrots, and beets are excellent choices because they are high in calories and nutrients, and they store well in cool, dry conditions.
- **Leafy Greens**: Spinach, kale, and lettuce grow quickly and can be harvested multiple times throughout the season. They provide essential vitamins and minerals.
- **Legumes**: Beans and peas are rich in protein and can be dried for long-term storage.

- **Squash and Pumpkins**: These are high in calories and have a long shelf life when properly stored.
- **Herbs**: In addition to adding flavor to meals, herbs like basil, oregano, thyme, and mint can have medicinal uses.

Maximizing Space

If you don't have much land, consider alternative gardening methods that allow you to grow more food in a small space:

- **Vertical Gardening**: Use trellises, hanging pots, or stacked planters to grow plants vertically, maximizing the use of limited space.
- **Container Gardening**: If you don't have access to soil or live in an urban environment, grow your crops in containers. This method allows you to move plants easily and control soil conditions.
- **Aquaponics or Hydroponics**: For those with the resources, setting up an **aquaponic** or **hydroponic** system can provide a continuous supply of fresh vegetables without the need for traditional soil-based gardening.

Growing Year-Round

Depending on your climate, growing food year-round may require some additional effort. **Greenhouses** and **cold frames** can extend your growing season, protecting crops from frost and cold weather. In colder months, you can also focus on **indoor gardening** using grow lights or sunny windows to grow leafy greens and herbs.

Preserving Food for Long-Term Storage

Once you begin harvesting food from your garden, it's essential to have a system in place for preserving it. Navy SEALs rely on compact, long-lasting food sources during missions, and similarly, preserving your food ensures you can store surplus produce for future use.

Canning

Canning is one of the most effective ways to preserve vegetables, fruits, and even meat for long-term storage. There are two main methods of canning:

- **Water Bath Canning**: Suitable for high-acid foods like tomatoes, fruits, and pickles. The food is packed into jars and submerged in boiling water to create a vacuum seal.
- **Pressure Canning**: For low-acid foods like vegetables, meats, and soups, a **pressure canner** is required to heat the jars to a high enough temperature to kill bacteria.

Dehydration

Dehydrating removes moisture from food, which prevents bacteria from growing and spoiling it. You can dehydrate vegetables, fruits, herbs, and even meat to make jerky. Dehydrated food is lightweight and compact, making it easy to store for long periods.

- **Electric Dehydrators**: These provide a reliable way to dry food at home.
- **Sun Drying**: If you live in a hot, dry climate, you can also sun-dry fruits and vegetables by placing them on racks in direct sunlight.

Fermentation

Fermenting foods not only preserves them but also enhances their nutritional value by introducing beneficial bacteria. Fermented foods like **sauerkraut**, **kimchi**, and **pickles** can be stored for months and provide important probiotics for gut health.

- **Lacto-Fermentation**: This process uses **salt** to preserve vegetables. Simply pack vegetables tightly in a jar, cover with brine, and let them ferment at room temperature for several days.

Freezing

While freezing is an option for preserving food, it depends on having a reliable source of power. If you have access to solar power or a generator, freezing can be an effective way to store surplus food for months.

16.3 - Sustainable Water Management: Ensuring a Continuous Supply of Clean Water

Just as important as food production is securing a **sustainable source of clean water**. Water is essential not only for drinking but also for cooking, sanitation, and growing food. In a long-term crisis, relying on stored water will only get you so far. Developing a renewable water system is crucial for long-term survival.

Rainwater Harvesting Systems

One of the most effective ways to secure a renewable water source is through **rainwater harvesting**. SEALs operating in remote environments often rely on natural water sources, and you can do the same by collecting rainwater from your roof or other surfaces.

Setting Up a Rainwater Collection System

To set up a basic rainwater harvesting system:

- **Gutters and Downspouts**: If your home has gutters, install a system that directs rainwater into storage barrels or tanks. Clean gutters regularly to ensure they are free of debris.
- **Rain Barrels**: Use **rain barrels** or larger **storage tanks** to collect water. Make sure they are made of food-grade materials and that they

are covered to prevent contamination from insects or debris.
- **First Flush Diverters**: Install a **first flush diverter** to redirect the first few minutes of rainfall away from the storage tank, as this water may contain dust, dirt, or bird droppings.

Filtering and Purifying Rainwater

While rainwater is generally safe, it may contain contaminants, especially if it has run off from roofs or other surfaces. Always **filter** and **purify** harvested rainwater before drinking it:

- **Pre-Filters**: Install a simple mesh filter to remove leaves, twigs, and debris before the water enters your storage tank.
- **Carbon Filters**: Use **carbon filters** to remove chemicals and improve the taste of water.
- **Boiling**: Boiling water is one of the most effective ways to kill bacteria, viruses, and parasites.
- **UV Sterilization**: **UV light systems** can be used to purify water by killing microorganisms.
- **Chemical Treatment**: In a pinch, you can use **water purification tablets** (such as iodine or chlorine dioxide) to disinfect water.

Alternative Water Sources

In addition to rainwater harvesting, consider other **alternative water sources** that may be available in your area:

- **Wells**: If you live in a rural area, consider drilling a well for access to groundwater. Wells provide a continuous supply of water but may require a hand pump or solar-powered pump in case of power outages.
- **Streams or Lakes**: If you have access to a nearby stream or lake, you can use it as a water source, but make sure to filter and purify the water before drinking.
- **Desalination**: For those living near the coast, **desalination** systems can convert seawater into drinkable water. Small-scale solar desalination units are available for personal use.

16.4 - Energy Independence: Powering Your Home Without the Grid

In a long-term crisis, **energy** becomes one of the most valuable resources. If the power grid goes down, you'll need alternative ways to power your home, run essential appliances, and keep your household comfortable. Navy SEALs are trained to operate in environments with no access to modern conveniences, often relying on portable generators and solar panels. You can apply similar principles to create an energy-independent home.

Solar Power: The Best Long-Term Energy Solution

Solar power is one of the most reliable and sustainable ways to generate electricity during a crisis. Whether you're powering your entire home or just essential devices, solar power allows you to harness the sun's energy and store it for later use.

Setting Up Solar Panels

To set up a solar power system:

- **Solar Panels**: Install **roof-mounted solar panels** or set up **portable solar panels** in your yard. Roof-mounted systems are more permanent and can generate significant amounts of power, while portable panels are more flexible and can be moved as needed.
- **Battery Storage**: To ensure you have power at night or on cloudy days, invest in a **solar battery storage system**. Batteries like the **Tesla Powerwall** or **deep cycle batteries** store excess energy generated during the day for use later.
- **Inverters and Charge Controllers**: You'll need an **inverter** to convert the direct current (DC) generated by the solar panels into alternating current (AC) for household use. A **charge controller** helps regulate the energy going into the batteries to prevent overcharging.

Powering Essential Devices

Even if you can't power your entire home, solar panels can be used to power **essential devices**:

- **Lighting**: Use solar power to keep lights on at night, improving security and comfort.
- **Communications**: Keep radios, phones, and other communication devices charged with solar energy.
- **Medical Devices**: If anyone in your household relies on **medical devices** (such as oxygen concentrators or CPAP machines), make sure you have enough solar capacity to power these critical devices.

Backup Generators: A Temporary Power Solution

While solar power is ideal for long-term sustainability, having a **backup generator** can provide additional security. SEALs often use **portable generators** in the field to power essential equipment, and having one at home ensures you have power during emergencies.

Types of Generators

- **Gasoline Generators**: These are the most common type of generator and can provide power for essential appliances like refrigerators, lights, and heaters. However, they require a steady supply of fuel, which may be hard to come by in a long-term crisis.

- **Propane Generators**: **Propane-powered generators** are more efficient than gasoline models and can run for longer periods. Propane is also easier to store for extended periods.
- **Solar Generators**: A **solar generator** uses solar panels to generate electricity and store it in a battery. These are quieter and more environmentally friendly than gas-powered generators, but they may not provide as much power.

Managing Fuel and Power Usage

To conserve fuel and power:

- **Prioritize Essential Appliances**: Use the generator only for critical appliances like refrigerators, heaters, or medical devices. Avoid running non-essential devices.
- **Rotate Power Usage**: If you have limited fuel, consider rotating power usage by running the generator only during certain times of day.

16.5 - Mental and Emotional Resilience: Thriving Through Psychological Endurance

In a long-term survival scenario, **mental and emotional endurance** can be just as important as physical resources. Navy SEALs are trained to withstand extreme stress, isolation, and uncertainty, developing the mental toughness needed to complete their missions no matter how difficult the circumstances. By adopting similar mental strategies, you can maintain **psychological resilience** throughout an extended crisis.

Maintaining Routine and Structure

One of the biggest challenges in a long-term crisis is the breakdown of **routine** and **structure**. Without the normal rhythms of daily life—work, social interactions, and outside obligations—time can begin to blur, leading to **disorientation**, **boredom**, and **anxiety**. SEALs maintain strict routines, even in isolated environments, to keep their minds sharp and their spirits focused.

Creating a Daily Schedule

Even in a crisis, create a **daily schedule** that includes regular tasks and activities:

- **Morning Routine**: Start the day with a structured morning routine that includes basic hygiene, exercise, and planning for the day ahead.
- **Work Tasks**: Assign specific tasks for the day, such as gardening, securing your perimeter, managing supplies, or performing maintenance tasks.

- **Exercise**: Physical activity is essential for both mental and physical well-being. Include regular exercise in your daily routine to reduce stress and improve overall health.
- **Rest and Relaxation**: Don't forget to schedule downtime. Rest is critical for mental clarity and preventing burnout.

Building Community Support

As discussed in **Chapter 14**, building a **community network** is critical for long-term survival. But beyond resource-sharing and security, a community provides **emotional and social support**, helping combat the loneliness and isolation that often come with extended crises.

- **Check-Ins**: Schedule regular check-ins with neighbors or members of your community network. Even a brief conversation can boost morale and remind you that you're not alone.
- **Collaborative Projects**: Working on group projects—such as building a greenhouse, setting up a water system, or organizing community security—fosters a sense of shared purpose and camaraderie.

Staying Motivated and Setting Goals

Long-term survival can sometimes feel like an endless slog, especially when there's no clear end in sight. SEALs are taught to **set small goals** to stay motivated during long missions, celebrating each victory no matter how small.

- **Set Daily and Weekly Goals**: Focus on short-term goals that can be achieved within a day or week. Whether it's planting a new crop, finishing a home repair, or learning a new skill, these small victories provide a sense of progress.
- **Celebrate Milestones**: Take time to **celebrate milestones**, even if they seem minor. Recognizing accomplishments boosts morale and reinforces the feeling of forward momentum.

Managing Stress and Anxiety

Stress and anxiety are inevitable in long-term crises, but it's important to manage these feelings to prevent **burnout** and **mental fatigue**.

- **Mindfulness and Meditation**: Practicing **mindfulness** or **meditation** helps reduce stress by bringing your focus back to the present moment. These techniques are often used by SEALs to manage stress in high-pressure environments.
- **Journaling**: Writing about your thoughts, feelings, and experiences can provide an

emotional outlet and help you process the challenges of long-term survival.

16.6 - Final Thoughts: Thriving Through Sustainability and Adaptability

Long-term survival requires more than just a well-stocked pantry or a fortified home. It demands **adaptability**, **resourcefulness**, and the ability to **build sustainable systems** that allow you to thrive in the face of uncertainty. Just as Navy SEALs must constantly adjust their strategies and maintain mental resilience during prolonged missions, you must learn to shift from short-term survival tactics to long-term sustainability.

By focusing on renewable resources—such as food production, water harvesting, and solar power—you can create a home environment that supports **self-sufficiency** for as long as needed. And by cultivating **mental and emotional resilience**, you can maintain the psychological endurance needed to stay strong throughout an extended crisis.

Remember, the goal is not just to survive but to create a life that allows you and your family to **thrive**, no matter how long the crisis lasts.

Chapter 17: Adapting to Changing Threats and Environmental Conditions

In a long-term crisis, nothing stays the same. Threats evolve, the environment changes, and the unexpected becomes the norm. Flexibility and adaptability are crucial survival traits. Navy SEALs train extensively in fluid tactical thinking, always ready to pivot strategies in the face of new challenges. In a prolonged bug-in scenario, you will face a range of changing conditions—shifting security risks, natural disasters, resource scarcity, and more. This chapter focuses on building the **flexibility** to adapt to changing environments and **the agility to adjust your tactics** to ensure you and your family stay safe and resilient.

The key to thriving in a long-term crisis is the ability to **see changes coming** before they overwhelm you and to **react quickly and efficiently** to new challenges. From fortifying your home against emerging threats to adjusting your supply management and security protocols, adaptability will be your most valuable asset.

17.1 - Identifying and Understanding New Threats

As a crisis stretches on, the threats you face will likely shift. Early in the crisis, security risks might involve opportunistic looters or desperate individuals. Over time, more organized groups or even hostile environmental conditions could become more significant concerns. SEALs are trained to constantly monitor and reassess their environment, scanning for emerging threats and adjusting their response plans accordingly. You must develop this same awareness to ensure your survival over the long term.

Assessing Evolving Security Risks

In the early stages of a crisis, the primary security threats often come from those seeking immediate resources—food, water, and supplies. As the crisis drags on, however, **security risks** may become more organized and dangerous. It's essential to continually assess the changing landscape and adjust your home defenses and security protocols accordingly.

Signs of Escalating Security Threats

- **Increased Group Activity**: As time progresses, you may notice an increase in **organized groups** operating in your area. These groups may be looking for resources, attempting to take control of communities, or even exploiting the chaos for criminal purposes.
- **Greater Desperation**: As resources become more scarce, people's desperation may grow. This can lead to more aggressive behavior,

including violent confrontations, home invasions, and opportunistic attacks.
- **Changes in Local Dynamics**: Keep an eye on the broader dynamics in your local area. If you notice a breakdown in law enforcement, government response, or community infrastructure, it's likely that the security situation will worsen over time.

Adjusting Your Security Strategy

Once you've identified signs of escalating threats, it's important to adjust your **security strategy** to reflect the new risks. SEALs are trained to constantly evolve their defensive tactics to match the threat level, and your bug-in security plan should be no different.

- **Fortify Perimeter Defenses**: Reinforce your home's **perimeter defenses**. This could mean adding additional layers of barriers, such as sandbags, spikes, or barbed wire. Increase the use of **motion sensors**, **security cameras**, and **lights** to detect threats early.
- **Create Backup Security Plans**: Develop **contingency plans** in case your primary defense system is breached. For example, set up **safe rooms** or **fallback positions** within your home where you and your family can retreat to if necessary.
- **Upgrade Weaponry and Training**: As threats become more organized and potentially violent, it

may be necessary to upgrade your defensive weapons. Ensure you have a diverse range of self-defense tools, from **firearms** to **non-lethal options** like pepper spray or tasers. Make sure everyone in your household is trained in their use.
- **Regular Patrols and Watch Schedules**: If you've formed a **community network** (as discussed in Chapter 14), organize regular security patrols and rotating watch schedules to ensure constant vigilance.

Monitoring New Environmental Challenges

Beyond human threats, environmental challenges may arise or worsen during a long-term crisis. Whether due to extreme weather, natural disasters, or secondary effects of the initial disaster (such as water contamination or infrastructure breakdowns), it's critical to be aware of changing environmental risks and adjust your survival plan accordingly.

Identifying Environmental Changes

- **Weather Shifts**: Climate patterns may become more extreme during a crisis, especially if the disaster is related to environmental events (such as a hurricane, earthquake, or wildfire). Pay attention to shifting weather patterns—extreme heat, cold, storms, or drought—and plan for how

they will impact your shelter, food production, and water supplies.
- **Natural Disasters**: In a prolonged crisis, natural disasters like **earthquakes**, **floods**, or **wildfires** may pose additional threats. Keep a close eye on weather forecasts and disaster reports to stay ahead of potential environmental risks.
- **Resource Contamination**: A long-term crisis may lead to **contaminated water**, **spoiled food**, or other environmental hazards. For example, damaged infrastructure or a lack of maintenance could lead to unsafe drinking water, while electrical outages may cause food spoilage. Be prepared to deal with contaminated resources by having **water filtration systems** and **alternative food sources** ready.

Adapting to Environmental Threats

- **Improve Shelter Insulation and Protection**: Ensure your home is well-insulated and protected against the elements. In cold climates, this may mean reinforcing windows, sealing drafts, and ensuring you have a reliable source of heat. In hot climates, focus on ventilation, shading, and cooling methods.
- **Prepare for Natural Disasters**: If you live in an area prone to natural disasters, make sure your home is fortified against these risks. For

example, use **sandbags** to protect against flooding, reinforce roofs to withstand heavy winds, and clear debris from around your home to reduce fire risk.
- **Establish Emergency Evacuation Routes**: Even in a long-term bug-in scenario, there may be situations where evacuation is necessary due to environmental hazards. Establish clear **evacuation routes** and keep **go-bags** packed with essentials in case you need to leave your home quickly.

17.2 - Resource Scarcity: Adapting Your Supply Management

As a long-term crisis progresses, the resources you initially stockpiled may begin to dwindle, and securing new supplies may become increasingly difficult. Navy SEALs are trained to manage resources efficiently and improvise when supplies run low, ensuring they can continue their mission despite limited provisions. You can apply these same principles to your own resource management, focusing on conservation, alternative supply sources, and efficient usage.

Conserving Supplies and Reducing Waste

In the early days of a crisis, it's common to rely on stored supplies and stockpiles. However, as the crisis continues, it becomes essential to **conserve** your resources and reduce waste to extend their longevity.

Food and Water Conservation

- **Rationing**: Implement a **rationing system** that carefully controls how much food and water each person consumes per day. Adjust portion sizes based on your remaining supplies and aim to stretch your resources as long as possible.
- **Avoiding Waste**: Be vigilant about avoiding food waste. Plan meals that use every part of an ingredient, including scraps. For example, use vegetable peelings to make broth or repurpose leftovers into new dishes.
- **Recycling Water**: In a resource-scarce environment, reusing water becomes essential. For example, you can **reuse greywater** (such as water used for washing dishes or clothes) for non-potable uses like watering plants or flushing toilets. Make sure to properly filter and purify any water before drinking it.

Energy and Fuel Conservation

- **Reducing Energy Consumption**: Be mindful of your energy usage by turning off lights, unplugging devices, and reducing the number of appliances you run simultaneously. Prioritize using power for essential activities, such as cooking, refrigeration, and communication.
- **Alternative Energy Sources**: If fuel supplies are running low, consider alternative energy sources, such as **solar power**, **wind energy**, or

hand-cranked devices for small electronics. Having multiple energy options ensures you're not reliant on a single power source.

Securing Alternative Supply Sources

As traditional supply chains break down and resources become more scarce, it's critical to develop **alternative supply sources** that allow you to replenish your stockpiles. SEALs often improvise by sourcing local materials and utilizing available resources in the field. You can apply this same approach by tapping into local, renewable resources.

Foraging and Hunting

If food supplies are running low, foraging for **wild edibles** and **hunting small game** can provide an alternative source of nutrition. Familiarize yourself with local edible plants, mushrooms, berries, and herbs, as well as areas where you can legally hunt or fish.

- **Foraging**: Learn to identify local wild edibles that are safe to consume, such as dandelions, wild greens, nuts, and berries. Always double-check for poisonous lookalikes, and be cautious about the environmental impact of foraging.
- **Fishing and Hunting**: If you have access to water sources, consider fishing to supplement your food supply. Hunting small game like rabbits, squirrels, or birds can provide protein. Be sure to familiarize yourself with local hunting

laws and ethical practices, and have the necessary gear on hand.

Bartering and Trade

If you've developed a **community network**, bartering and trading can be an effective way to acquire new supplies. SEAL teams often rely on local trade to secure needed resources, and you can do the same by exchanging surplus items with neighbors or others in your community.

- **Identify Trade Goods**: Determine which items or skills you have in surplus that might be valuable to others. This could include extra food, tools, fuel, or even knowledge (such as medical skills or carpentry).
- **Organize a Barter Market**: If your community network is strong, consider organizing a **barter market** where individuals can come together to trade goods and services. This allows everyone to fill gaps in their supplies without relying on outside sources.

17.3 - Flexibility in Shelter: Adjusting Your Living Environment

Over time, your home may need to adapt to changing environmental conditions, security risks, or resource availability. SEALs are trained to build flexible shelters that can be easily adjusted or relocated based on changing threats. Similarly, your bug-in shelter should be able to evolve with the crisis, providing you with protection, comfort, and sustainability over the long term.

Expanding or Modifying Your Shelter

As the crisis evolves, you may need to **expand** or **modify** your shelter to accommodate new challenges. For example, if you anticipate increased security threats, you may need to reinforce your home's defenses. If you're producing more food, you may need to build additional storage or create outdoor structures like greenhouses or water collection systems.

Adding Outdoor Structures

If your home has outdoor space, consider adding structures that increase your ability to produce food, store resources, or protect your family from environmental threats.

- **Greenhouses**: Build a small greenhouse to extend your growing season and protect plants from harsh weather. This allows you to grow fresh food year-round.
- **Storage Sheds**: If you've stockpiled resources or are producing surplus food, you'll need additional storage space. A well-insulated shed

can store tools, seeds, fuel, and preserved food safely.
- **Rainwater Collection Systems**: Install a larger rainwater collection system if you find that your current setup isn't providing enough water for your household's needs. Make sure to add filtration and purification systems to ensure the water is safe to drink.

Reinforcing Your Home's Defenses

As discussed earlier, you may need to upgrade your home's defenses to deal with escalating security threats. SEALs often fortify their positions based on the intensity of enemy threats, and your home should be able to evolve in response to new security challenges.

- **Boarding Windows**: In high-risk situations, board up vulnerable windows or use **metal security bars** to prevent forced entry.
- **Hidden Storage**: Create **hidden compartments** within your home to store valuables, food, or weapons, keeping them safe from intruders.

Relocating or Retreating

In some cases, it may become necessary to **relocate** or **retreat** from your home temporarily. SEALs are taught to assess the risk of staying in one place for too long and to be prepared to relocate if the situation becomes untenable.

Bug-Out Locations

Even if you're committed to bugging in, it's wise to have a **bug-out location** as a backup plan. This could be a cabin in a rural area, a friend's home, or even a temporary shelter outside the city. Ensure your bug-out location is fully stocked with essentials and that you have a clear route for getting there.

Go-Bags

Have **go-bags** packed for each member of your household in case you need to leave quickly. These should contain:

- **Food and Water**: Enough for 72 hours.
- **Clothing and Shelter**: Warm clothes, a sleeping bag, and a tarp or tent.
- **First Aid Kit**: Including any necessary prescription medications.
- **Self-Defense Tools**: Pepper spray, a firearm, or other protective equipment.
- **Important Documents**: Copies of ID, insurance, and property deeds in a waterproof container.

17.4 - Adapting Your Mindset: Embracing Change and Overcoming Fatigue

Long-term crises require **mental agility** and the ability to cope with constant change. SEALs are taught to expect the unexpected and to mentally prepare for mission parameters that shift at a moment's notice. You'll need to develop a similar mindset, one that embraces flexibility, continuous learning, and resilience in the face of the unknown.

Dealing with Fatigue and Burnout

In any long-term survival situation, **fatigue** and **burnout** are major risks. Over time, the stress of maintaining your home, securing resources, and protecting your family can lead to mental exhaustion. SEALs are trained to manage burnout through physical and mental recovery techniques, and you can apply these same principles to your own survival.

Building Recovery Periods into Your Routine

To avoid burnout, it's essential to **schedule regular recovery periods**. Even during a crisis, you need time to rest, recuperate, and regain your mental clarity.

- **Sleep**: Ensure you're getting enough sleep, even if it means setting up rotating watch schedules with other household members.
- **Micro-Recovery**: Take short breaks throughout the day to rest your mind and body. These micro-recovery periods can involve simple activities like sitting quietly, meditating, or listening to music.
- **Physical Activity**: Engage in regular **physical exercise**, such as walking, stretching, or

bodyweight exercises. Physical activity helps release stress and maintain your overall health.

Flexibility in Decision-Making

One of the hallmarks of Navy SEAL training is the ability to make **fast, effective decisions** in fluid and unpredictable situations. Long-term survival requires a similar approach, as you will constantly need to reassess your situation and make adjustments based on new information.

Adjusting Strategies

As the crisis evolves, be prepared to **change strategies** as needed. For example, if a particular source of food or water becomes unreliable, pivot to a new method of securing those resources. Be willing to experiment with new approaches, even if they weren't part of your original plan.

Learning from Mistakes

Not every decision will be perfect. Be willing to **learn from mistakes** and adjust your course accordingly. Rather than becoming discouraged, treat setbacks as learning opportunities that will make you more resilient in the future.

17.5 - Final Thoughts: Flexibility as a Survival Tool

In a long-term crisis, **flexibility** is one of the most important tools you have. The ability to adapt your strategies, resources, and mindset to meet evolving threats and changing environmental conditions will determine your survival over the long term. Just as Navy SEALs constantly reassess their environment and adjust their tactics, you too must embrace a mindset of continuous adaptation.

By learning to recognize new threats, securing alternative resources, adjusting your living environment, and cultivating mental resilience, you can thrive in even the most uncertain and challenging situations. The world may change around you, but your ability to **stay agile and resourceful** will ensure your household remains strong, safe, and self-sufficient.

Chapter 18: Strategic Leadership and Decision-Making in a Long-Term Crisis

In a long-term crisis, strong leadership can mean the difference between chaos and order, failure and survival. Leadership in a survival scenario goes beyond just managing resources and directing efforts; it requires you to inspire, motivate, and guide others through periods of uncertainty and fear. The ability to make sound decisions quickly, communicate effectively, and stay calm under pressure are skills that Navy SEALs develop through rigorous training, and they are essential for anyone leading a group or household in a crisis.

This chapter will provide an in-depth guide to **strategic leadership**, covering topics such as **decision-making in high-pressure environments**, **maintaining group morale**, and **navigating conflicts**. By applying the principles of SEAL leadership to your crisis preparedness strategy, you can build the skills necessary to lead with confidence and resilience.

18.1 - The Role of a Leader in a Survival Scenario

In a long-term survival situation, the role of a leader is to **provide direction, maintain order**, and **ensure the group's well-being**. Leadership doesn't necessarily mean being the loudest or most authoritative person in the room—it's about being the person who others trust to make the right decisions in critical moments.

Key Responsibilities of a Leader

As a leader in a crisis, your responsibilities will encompass several areas:

- **Decision-Making**: You must be prepared to make **tough decisions**, often with incomplete information, under pressure, and in rapidly changing circumstances.
- **Resource Management**: Ensuring that the group has enough food, water, medical supplies, and other essentials is one of the most important duties of a leader in a survival scenario.
- **Security and Protection**: A leader must ensure the physical safety of the group, whether that means organizing defenses, developing a security plan, or protecting the group from external threats.
- **Morale and Motivation**: Emotional resilience is just as important as physical preparedness. A good leader maintains **group morale**, keeps

people focused, and prevents panic or despair from spreading.

Traits of an Effective Crisis Leader

The most successful leaders in survival situations exhibit several key traits. Navy SEALs are trained to embody these characteristics in their missions, and you can develop these qualities to lead your household or group through any crisis:

- **Calm Under Pressure**: Even in the face of danger or uncertainty, a leader must remain calm and composed. Panic or indecision can spread quickly, so it's crucial to project a sense of confidence and control.
- **Adaptability**: In a long-term crisis, things rarely go as planned. A good leader is flexible and able to adapt strategies and decisions to meet changing circumstances.
- **Decisiveness**: Leaders must be able to make decisions quickly and effectively, especially in high-pressure environments. Hesitation or procrastination can lead to missed opportunities or increased risk.
- **Empathy**: A strong leader understands the emotional and psychological needs of the group. Showing compassion and empathy builds trust and strengthens the group's unity.
- **Communication**: Clear and direct communication is vital in a crisis. Leaders must

ensure that everyone understands the plan, their role, and what is expected of them.

18.2 - Decision-Making Under Pressure

In a crisis, every decision you make has consequences, and some decisions may have life-or-death implications. SEALs train extensively in **high-pressure decision-making**, often in environments where they have limited time and incomplete information. Adopting a systematic approach to decision-making can help you stay focused and make the best possible choices, even when the stakes are high.

The OODA Loop: A Framework for Decision-Making

One of the most effective models for decision-making under pressure is the **OODA Loop**. Developed by military strategist John Boyd, the OODA Loop stands for:

- **Observe**: Gather as much information as you can from your environment. What are the immediate threats? What resources are available? Who is affected by the decision?
- **Orient**: Analyze the information and assess how it relates to the current situation. Consider the possible outcomes of various courses of action.
- **Decide**: Based on your analysis, choose a course of action that addresses the situation effectively.

- **Act**: Implement your decision quickly and decisively, and then monitor the outcome to adjust as necessary.

SEALs are trained to cycle through the OODA Loop rapidly, constantly adjusting their plans and strategies as new information becomes available. In a survival scenario, the ability to **make decisions quickly** and then adapt based on feedback is essential for staying ahead of threats and challenges.

Balancing Risk and Reward

When making decisions in a crisis, you often have to weigh the **risks** against the **potential benefits**. Navy SEALs are taught to perform rapid **risk assessments**, evaluating the likelihood of success against the potential for failure or danger. As a leader, you'll need to do the same.

- **Immediate vs. Long-Term Risk**: Some decisions may have immediate benefits but could create long-term risks. For example, using up a large portion of your food supply to boost morale in the short term might leave you vulnerable to starvation later. Always consider the long-term impact of each decision.
- **Cost-Benefit Analysis**: Evaluate the resources needed for each course of action against the potential benefits. For instance, sending a team out to scout for new supplies might yield critical

resources but could also expose the group to security risks. Weigh the importance of new supplies against the danger of losing key members of your group.

Trusting Your Instincts

While it's important to analyze your options carefully, in many situations you won't have the luxury of time. SEALs are trained to **trust their instincts** and make decisions based on their gut feeling when time is of the essence. In a survival scenario, you may need to act quickly, even without perfect information. Developing **good judgment** over time by observing your environment and understanding human behavior will allow you to make more intuitive decisions in high-stress situations.

18.3 - Maintaining Group Morale and Unity

In a long-term crisis, one of the greatest challenges is maintaining **group morale**. Even the most well-prepared groups can be torn apart by internal conflicts, fear, and despair if they don't have a strong leader to guide them through difficult times. SEALs operate in some of the most stressful environments on earth, and maintaining mental and emotional resilience is crucial to their success. Similarly, as a leader in a survival scenario, it's your job to keep your group motivated, focused, and unified.

The Importance of Purpose

One of the most powerful tools for maintaining morale is giving people a **sense of purpose**. SEALs are able to endure extreme physical and psychological challenges because they believe in the **mission**. In a survival scenario, the "mission" is the group's survival, but people need to be reminded of this higher goal in order to stay motivated.

Creating a Shared Goal

- **Clear Objectives**: Ensure that everyone understands the **shared goal**—surviving the crisis. Break this larger goal down into smaller, more manageable objectives, such as securing water, building a shelter, or planting a garden. Clear, actionable goals help keep people focused.
- **Daily Tasks**: Assign each member of the group specific tasks that contribute to the overall survival plan. When people feel that they are contributing, it boosts their sense of purpose and self-worth.
- **Involvement in Decision-Making**: Involve key members of the group in the **decision-making process**. This not only distributes leadership but also helps build investment in the group's success.

Managing Stress and Anxiety

Crises are inherently stressful, and stress can easily spiral into panic or depression if it's not managed properly. As a leader, it's your responsibility to help your group manage their stress levels and **maintain emotional resilience**.

Signs of Stress and Burnout

Be on the lookout for signs that someone in the group may be experiencing burnout or severe stress, including:

- **Irritability** or **anger**
- **Withdrawal** from group activities or social interaction
- **Lack of concentration** or **poor decision-making**
- **Physical symptoms** such as headaches, fatigue, or sleep problems

When you notice these signs, address them early. Sometimes a simple conversation or offering to lighten someone's workload can make a big difference.

Promoting Rest and Recovery

Ensure that everyone in the group, including yourself, gets adequate **rest** and **recovery**. SEALs understand the importance of downtime, even in the midst of high-intensity missions. Incorporating regular periods of rest, relaxation, and physical recovery into your group's daily routine can prevent burnout.

- **Rotating Watch Shifts**: If your group is responsible for security, make sure that people rotate shifts so that no one is overworked.
- **Scheduled Breaks**: Allow for downtime where people can rest, sleep, or engage in low-stress activities. This could include reading, playing a game, or having quiet moments of reflection.
- **Physical Activity**: Encourage regular **physical activity**, which can help reduce stress and boost morale. Even something as simple as a group walk or light exercise can improve mental and physical health.

Conflict Resolution and Group Cohesion

No matter how united a group is, **conflicts** will arise, especially in a stressful survival scenario. SEALs train in **team dynamics** to prevent small disagreements from spiraling into larger conflicts. As a leader, you must be proactive in resolving disputes before they undermine group cohesion.

Addressing Conflict Early

- **Don't Avoid Issues**: Address conflicts as soon as they arise. Allowing grievances to fester will only cause more problems later.

- **Mediation**: Act as a mediator between conflicting parties. Encourage open, respectful communication and ensure that both sides are heard.
- **Problem-Solving Approach**: Focus on **problem-solving** rather than blame. Instead of dwelling on what went wrong, direct the group toward finding solutions that benefit everyone.

Building Unity Through Shared Experiences

Strengthen group cohesion by creating **shared experiences**. SEALs build trust through shared challenges, and you can create similar bonds within your group through teamwork and cooperation.

- **Group Projects**: Engage in projects that require cooperation, such as building a shelter, organizing a food supply, or setting up security systems. These activities help build trust and a sense of camaraderie.
- **Celebrate Milestones**: Acknowledge and celebrate small successes as a group. Whether it's successfully defending the home or harvesting the first crops from a garden, celebrating achievements boosts morale and fosters unity.

18.4 - Navigating Leadership Challenges in a Crisis

Leading a group in a long-term crisis comes with numerous challenges, from managing **scarce resources** to balancing **competing needs** within the group. SEALs are often placed in leadership roles where they must make difficult decisions, sometimes choosing between competing priorities. Navigating these challenges requires strong **emotional intelligence**, clear **communication**, and the ability to make **difficult choices** under pressure.

Balancing Individual and Group Needs

In a survival scenario, there will be times when individual needs conflict with the needs of the group. SEALs are trained to prioritize the mission over personal interests, but this can be difficult to enforce in a civilian setting where emotions and personal relationships play a larger role.

Setting Group Priorities

- **Safety First**: In any crisis, safety should be the number one priority. Make it clear that decisions will always be made with the group's overall safety in mind.
- **Resource Allocation**: When it comes to distributing limited resources, such as food, water, or medical supplies, prioritize based on

need. For example, those who are ill, injured, or performing strenuous tasks may require more resources.
- **Open Discussions**: Encourage **open discussions** about resource management and group priorities. By including others in the conversation, you reduce resentment and build understanding about why certain decisions are made.

Handling Difficult Decisions

In a crisis, you may be faced with making decisions that are unpopular or difficult for the group to accept. SEAL leaders are trained to make tough choices under pressure, and you will need to do the same. Whether it's enforcing a rationing system, deciding when to relocate, or determining how to deal with security threats, difficult decisions must be made for the greater good.

Clear Communication in Decision-Making

- **Explain the Rationale**: When making tough decisions, explain your reasoning clearly. People are more likely to accept difficult choices if they understand why they are being made.
- **Stand Firm**: Be prepared to stand by your decisions, even if they are unpopular. In a survival scenario, the leader's responsibility is to prioritize the group's safety and well-being, which may require making hard choices.

Managing Leadership Fatigue

Leadership itself can be exhausting, especially in a long-term crisis. SEALs are taught to recognize the signs of **leadership fatigue** and to rely on their teams for support when needed. As a civilian leader, you will face similar challenges and must learn to manage your own stress levels to remain an effective leader.

Delegating Responsibility

- **Empowering Others**: Don't be afraid to **delegate responsibilities** to others within the group. By sharing the workload, you reduce your own stress and empower others to take on leadership roles.
- **Developing New Leaders**: Identify potential leaders within your group and help develop their skills. This not only reduces the burden on you but also ensures that there are other capable individuals who can step in if needed.

Taking Breaks

- **Rest and Recharge**: Even the best leaders need time to rest and recharge. Make sure you schedule time for yourself to rest, reflect, and recover. Taking care of your own well-being allows you to be a more effective leader over the long term.

18.5 - Final Thoughts: Leadership as a Lifeline in Long-Term Survival

Leadership is the cornerstone of successful long-term survival. In a crisis, people look to their leader for guidance, support, and direction. By adopting the leadership principles used by Navy SEALs—calm under pressure, adaptability, empathy, and decisiveness—you can guide your household or group through the most challenging situations.

Remember, leadership is not just about making decisions; it's about **inspiring confidence**, **building trust**, and creating a sense of **purpose** in those around you. By keeping your group united, motivated, and focused on the ultimate goal of survival, you will provide the foundation they need to overcome any obstacles that arise.

In the end, strong leadership can turn an uncertain and dangerous situation into one where your group thrives, not just survives.

Chapter 19: Advanced Medical Care and First Aid: Self-Reliance in Long-Term Survival

In a long-term survival scenario, having the knowledge and skills to provide **advanced medical care** is critical. When hospitals are inaccessible, and professional medical help is unavailable, your ability to treat injuries and illnesses could be the difference between life and death for yourself or a loved one. While basic first aid is essential, a prolonged crisis demands a more advanced understanding of **wound care**, **infection prevention**, **trauma treatment**, and **long-term health management**.

Navy SEALs undergo extensive training in battlefield medicine, learning how to stabilize injuries, prevent infection, and keep their team members alive in extreme conditions. This chapter will teach you how to adapt these life-saving techniques for use in a bug-in scenario, covering both emergency treatments and ongoing health care needs in a survival situation.

19.1 - Building an Advanced Medical Kit for Long-Term Survival

One of the first steps in preparing for medical emergencies during a long-term crisis is assembling a comprehensive **medical kit**. While most people have basic first aid supplies like bandages and antiseptics, a survival scenario requires a much more robust medical kit capable of addressing a wide range of injuries and illnesses. This kit will need to be stocked with supplies for both **immediate care** and **long-term health maintenance**.

Basic First Aid Supplies

Your medical kit should contain all the essential first aid supplies needed for treating minor injuries, such as cuts, scrapes, burns, and sprains. These include:

- **Bandages and Dressings**: Sterile gauze pads, adhesive bandages, and medical tape for dressing wounds.
- **Antiseptics**: Alcohol wipes, hydrogen peroxide, iodine, or Betadine for cleaning wounds and preventing infection.
- **Tweezers and Scissors**: For removing debris from wounds or cutting bandages.
- **Pain Relievers**: Over-the-counter medications like ibuprofen, acetaminophen, and aspirin for pain relief and reducing inflammation.
- **Burn Ointment**: For treating minor burns and preventing infection.
- **Cold Compresses**: Instant cold packs for reducing swelling and treating sprains or strains.

Advanced Medical Supplies

In a long-term crisis, you'll need medical supplies that go beyond basic first aid. These supplies will help you treat more serious injuries, manage infections, and handle trauma care:

- **Sutures and Wound Closure Supplies**: Having the ability to **close wounds** is crucial in preventing infection and promoting healing. Sterile sutures, surgical needles, and a needle holder (or sterile skin adhesive strips like Steri-Strips) should be part of your kit.
- **Hemostatic Agents**: Products like **QuikClot** or **Celox** help stop severe bleeding in cases of deep wounds or arterial bleeding.
- **Tourniquet**: A **tourniquet** is essential for stopping life-threatening bleeding from limbs, particularly in cases of traumatic injury or amputations.
- **Splints and Braces**: To immobilize broken bones or sprained joints, include **splints**, SAM splints, or rigid braces in your kit.
- **Antibiotics**: Prescription **antibiotics** (such as amoxicillin, ciprofloxacin, or doxycycline) should be stored for treating bacterial infections, as access to a doctor for prescriptions may not be available during a crisis.
- **Sterile Gloves and Masks**: Always use **sterile gloves** and face masks when treating wounds to reduce the risk of infection.

- **IV Supplies and Fluids**: If you or someone in your group has medical training, it can be useful to include **IV bags** (such as saline or lactated Ringer's solution) and IV administration sets for rehydrating patients who are severely dehydrated or in shock.
- **Stethoscope and Blood Pressure Cuff**: Having basic diagnostic tools like a **stethoscope** and **blood pressure cuff** can help you monitor vital signs and detect potential complications early.
- **Emergency Blankets**: Hypothermia can set in after injury, especially if someone is in shock or has lost a lot of blood. Emergency **Mylar blankets** will help keep the patient warm and prevent hypothermia.

Long-Term Health Maintenance Supplies

For long-term survival, you'll also need to consider medical supplies that help maintain **overall health** and prevent chronic conditions from worsening. This is especially important if someone in your group has a pre-existing condition such as **diabetes**, **asthma**, or **heart disease**.

- **Prescription Medications**: Try to stockpile an extra supply of any necessary **prescription medications** like insulin, inhalers, or blood pressure medicine. Talk to your doctor about how to safely store and extend the shelf life of these medications.

- **Vitamins and Supplements**: In a survival situation, access to fresh fruits, vegetables, and other nutrient-dense foods may be limited. **Multivitamins** and essential supplements (such as vitamin D, calcium, or potassium) can help prevent deficiencies.
- **Diabetes Supplies**: If anyone in your group is diabetic, ensure you have an adequate supply of **insulin**, **blood sugar monitors**, **lancets**, and **glucose tablets**.
- **Asthma Inhalers**: Stock extra **rescue inhalers** and any other asthma medications if someone in your group has asthma.

19.2 - Treating Common Injuries in a Survival Scenario

In a long-term bug-in scenario, injuries are likely to happen, whether it's a cut from preparing firewood, a sprained ankle from moving heavy supplies, or a burn from cooking over an open flame. Knowing how to treat these common injuries effectively can prevent complications like infections or further damage, keeping your group safe and healthy.

Wound Care and Infection Prevention

One of the most critical aspects of survival medicine is **wound care**. In a survival scenario, even a minor cut can become life-threatening if it becomes infected. SEALs are trained to prioritize wound care in the field to prevent sepsis or gangrene, and you should adopt the same mindset in your home survival situation.

Step-by-Step Wound Treatment

1. **Stop the Bleeding**: Apply pressure to the wound using a sterile bandage or cloth to stop the bleeding. If the wound is large or involves severe bleeding, use a **tourniquet** or apply a **hemostatic agent** like QuikClot.
2. **Clean the Wound**: Once the bleeding is controlled, thoroughly clean the wound with **clean water** or a saline solution. If available, use antiseptic solutions like **hydrogen peroxide** or **Betadine** to disinfect the area.
3. **Remove Debris**: If there is dirt, gravel, or other debris in the wound, carefully remove it using **sterile tweezers**. Be sure to wash your hands or wear sterile gloves to avoid introducing bacteria into the wound.
4. **Apply Antibiotic Ointment**: Once the wound is clean, apply an **antibiotic ointment** like Neosporin to help prevent infection.
5. **Cover the Wound**: Use **sterile gauze** or an adhesive bandage to cover the wound and protect it from further contamination. Change the dressing daily or if it becomes wet or dirty.
6. **Monitor for Infection**: Signs of infection include **redness**, **swelling**, **warmth**, **pus**, and increasing **pain** around the wound. If infection

sets in, oral antibiotics may be needed to prevent sepsis.

Burns

Burns are common in survival situations, especially if you're cooking over an open flame or using alternative heat sources. Treating burns quickly and effectively can prevent infection and minimize scarring.

Step-by-Step Burn Treatment

1. **Cool the Burn**: Immediately run cool (but not cold) water over the burn for at least 10-15 minutes to reduce pain and prevent further damage to the skin.
2. **Do Not Use Ice**: Applying ice can cause further damage to the skin and worsen the burn.
3. **Apply Burn Ointment**: Once the burn has cooled, apply a **burn ointment** or **aloe vera gel** to soothe the skin and prevent infection.
4. **Cover the Burn**: Use a **sterile non-stick dressing** to cover the burn and protect it from infection. Avoid using adhesive bandages directly on the burn, as they may stick to the damaged skin.
5. **Pain Management**: Over-the-counter **pain relievers** like ibuprofen or acetaminophen can help reduce pain and swelling.
6. **Monitor for Infection**: Keep an eye on the burn for signs of infection, especially if the skin blisters

or breaks. If the burn becomes red, swollen, or starts to ooze pus, you may need to start an oral antibiotic.

Sprains and Fractures

In a survival situation, physical labor increases the risk of **sprains**, **strains**, and even **fractures**. Knowing how to immobilize injured limbs and reduce swelling can help prevent further damage.

Treating a Sprain

1. **Rest**: Have the injured person rest and avoid putting weight on the sprained joint.
2. **Ice**: Apply an **ice pack** or cold compress to the injured area for 15-20 minutes every hour to reduce swelling.
3. **Compression**: Wrap the injured joint with an **elastic bandage** to provide support and reduce swelling. Be careful not to wrap too tightly, as this can cut off circulation.
4. **Elevation**: Keep the injured limb elevated above the heart to reduce swelling.
5. **Pain Relief**: Over-the-counter pain relievers like ibuprofen or aspirin can help reduce pain and inflammation.

Treating a Fracture

1. **Immobilize the Limb**: Use a **splint** or **SAM splint** to immobilize the broken bone. A splint

can be made from rigid materials like wood, metal, or cardboard. Wrap the splint securely with bandages to keep the bone in place, but don't wrap so tightly that circulation is restricted.
2. **Prevent Shock**: If the person shows signs of **shock** (such as pale skin, rapid breathing, or confusion), lay them down with their legs elevated and keep them warm with a blanket.
3. **Seek Medical Help**: If possible, arrange to have the person transported to a medical facility for proper care. In a long-term survival scenario, managing a fracture may require more advanced intervention.

19.3 - Managing Chronic Illnesses in a Long-Term Crisis

In addition to emergency injuries, it's important to consider the management of **chronic illnesses** in a long-term crisis. For individuals with conditions like **diabetes**, **heart disease**, **asthma**, or **autoimmune disorders**, maintaining health can be a significant challenge when access to medication and professional care is limited. By planning ahead and taking preventive measures, you can reduce the risk of complications.

Diabetes Management

If someone in your group has **diabetes**, particularly **insulin-dependent diabetes**, it's essential to plan for how to manage their condition in a crisis.

Stockpiling Insulin and Supplies

- **Insulin**: Talk to a doctor about building up a stockpile of **insulin**, as this is a life-saving medication for diabetics. Proper storage is important, as insulin needs to be kept cool, ideally between 36°F and 46°F (2°C and 8°C). Consider investing in a **solar-powered refrigerator** or **insulated cooler** for long-term storage.
- **Glucose Monitors**: Keep extra **blood sugar monitors**, **test strips**, and **lancets** on hand to regularly check blood sugar levels.
- **Glucose Tablets**: Have **glucose tablets** or other fast-acting sugar sources to treat low blood sugar (hypoglycemia).

Dietary Adjustments

In a survival scenario, maintaining a **diabetic-friendly diet** may be difficult. Focus on stocking high-fiber, low-sugar foods like **beans**, **whole grains**, and **vegetables** to help stabilize blood sugar levels. Avoid high-sugar, processed foods that can cause spikes in blood sugar.

Asthma and Respiratory Conditions

For individuals with **asthma** or other respiratory conditions, **air quality** and **access to inhalers** are critical concerns.

Stockpiling Inhalers and Medication

- **Rescue Inhalers**: Ensure you have an adequate supply of **rescue inhalers** (albuterol) for immediate relief during asthma attacks.
- **Long-Term Control Medications**: If the person uses long-term control medications (such as corticosteroids), try to stockpile these as well.

Improving Air Quality

- **Air Filters**: If you're sheltering indoors for a long period, poor air quality from dust, mold, or smoke could exacerbate asthma symptoms. Consider using **portable air filters** or creating a **DIY air filtration system** with a box fan and a HEPA filter.
- **Humidity Control**: Use a **humidifier** or **dehumidifier** to maintain a healthy level of indoor humidity, which can help prevent asthma attacks.

Heart Disease and High Blood Pressure

For those with **heart disease** or **high blood pressure**, maintaining **cardiovascular health** in a crisis is a priority. Stress, poor diet, and lack of physical activity can worsen these conditions, so it's important to be proactive in managing heart health.

Medication Management

- **Stockpile Medications**: If the person is on **blood pressure medications**, **blood thinners**, or **cholesterol-lowering drugs**, ensure that you have an adequate supply to last through the crisis.
- **Monitor Blood Pressure**: Use a **blood pressure monitor** to regularly check blood pressure levels and ensure they are within a safe range.

Lifestyle Adjustments

- **Exercise**: Encourage regular **physical activity** to maintain cardiovascular health, even if it's just walking around the house or yard.
- **Diet**: Focus on a heart-healthy diet rich in whole grains, fruits, vegetables, and lean proteins. Avoid excessive salt and processed foods that can contribute to high blood pressure.

19.4 - Emergency Trauma Care and Triage

In a survival scenario, you may face situations where multiple people are injured at once, or where someone experiences a severe, life-threatening injury. In these cases, it's essential to understand the basics of **emergency trauma care** and **triage**, which is the process of prioritizing medical care based on the severity of injuries.

Performing Triage in a Survival Situation

When faced with multiple casualties, triage helps you decide who needs immediate medical attention and who can wait for treatment. The goal is to prioritize **life-threatening injuries** that require immediate intervention while stabilizing those with less severe injuries.

Triage Categories

- **Immediate (Red Tag)**: These individuals have life-threatening injuries that can be treated with immediate care. Examples include severe bleeding, chest trauma, or head injuries. Without prompt intervention, they will not survive.
- **Delayed (Yellow Tag)**: These individuals have serious but non-life-threatening injuries that require care but are not immediately fatal. Examples include broken bones or moderate burns.
- **Minor (Green Tag)**: These individuals have minor injuries that can be treated after more critical patients have been stabilized. Examples include small cuts or bruises.
- **Deceased (Black Tag)**: Unfortunately, in some cases, individuals may have injuries that are not survivable, or they may have already passed away. Focus your efforts on those who can still be saved.

Emergency Trauma Care Techniques

If someone in your group suffers a traumatic injury, knowing how to provide **immediate care** can save their life. Navy SEALs are trained to respond to trauma with speed and precision, focusing on stabilizing the patient and preventing further damage.

Managing Severe Bleeding

Stopping severe bleeding is the top priority in trauma care, as blood loss can quickly lead to shock and death.

1. **Apply Direct Pressure**: Use sterile gauze or a clean cloth to apply direct pressure to the wound. If the bleeding doesn't stop, add more layers of gauze and continue pressing firmly.
2. **Use a Tourniquet**: For severe bleeding from a limb, apply a **tourniquet** above the injury. Tighten the tourniquet until the bleeding stops, and note the time it was applied. Do not remove the tourniquet until the person is in a safe medical environment, as it can cause further damage if not handled properly.
3. **Apply a Hemostatic Agent**: If the bleeding is particularly difficult to control, use a **hemostatic agent** like QuikClot to help the blood clot and stop the bleeding.

Treating Head and Neck Injuries

Head and neck injuries are particularly dangerous and require immediate stabilization.

1. **Immobilize the Head and Neck**: If you suspect a neck or spine injury, do not move the person unless it's absolutely necessary. Stabilize their head and neck by placing rolled towels or clothing on either side to keep them from moving.
2. **Monitor for Concussion or Brain Injury**: Symptoms of a concussion or brain injury include confusion, memory loss, slurred speech, nausea, and loss of consciousness. If the person loses consciousness, monitor their breathing and be prepared to perform CPR if necessary.

Treating Chest Trauma

Injuries to the chest, such as broken ribs or a punctured lung, can lead to **breathing difficulties**. In severe cases, chest trauma can result in a **collapsed lung** (pneumothorax), which requires immediate attention.

1. **Seal Open Chest Wounds**: If there is a penetrating wound to the chest (such as a gunshot or stab wound), cover it with an airtight seal. Use plastic wrap or a chest seal to prevent air from entering the wound, which could worsen the injury.

2. **Monitor Breathing**: Keep the person sitting upright or lying on their injured side to make breathing easier. Watch for signs of respiratory distress, such as rapid, shallow breathing, or a bluish tint to the skin (cyanosis).

Treating Shock

After a severe injury, the body may go into **shock**, a life-threatening condition that occurs when blood pressure drops too low to sustain the body's vital organs.

1. **Lay the Person Flat**: If someone is showing signs of shock (pale skin, rapid breathing, confusion, or fainting), lay them flat on their back with their legs elevated about 12 inches.
2. **Keep Them Warm**: Cover the person with a blanket or emergency Mylar blanket to maintain body heat and prevent hypothermia.
3. **Do Not Give Food or Drink**: If the person is in shock, do not give them anything to eat or drink, as this could lead to choking.

19.5 - Final Thoughts: Building Confidence in Your Medical Skills

Medical emergencies can be overwhelming, especially in a survival scenario where help may not be immediately available. However, by building your **medical knowledge** and assembling a comprehensive

medical kit, you can become a more effective leader and caregiver in a crisis. Just as Navy SEALs train extensively to handle the most challenging environments, you can develop the skills and confidence needed to provide medical care for your family or group in a prolonged survival situation.

Remember, preparation is key. Invest time in learning **first aid**, **wound care**, and **emergency trauma treatment**. By doing so, you'll be equipped to handle the unexpected and increase the chances of survival for those in your care.

Chapter 20: Mental and Emotional Resilience: Thriving in a Long-Term Crisis

In the chaos of a long-term crisis, the ability to maintain **mental and emotional stability** becomes as important as having food, water, and shelter. The challenges you face won't just be physical—they will also take a heavy toll on your mind and spirit. Stress, fear, and uncertainty can quickly lead to mental exhaustion, burnout, and emotional breakdowns. In the military, Navy SEALs endure rigorous psychological training to ensure they remain resilient, adaptable, and focused under extreme conditions. As you prepare to bug in for an extended period, developing **mental resilience** is critical for survival.

This chapter will help you build the mindset necessary to stay strong and lead others through periods of fear, isolation, and emotional strain. You'll learn to overcome the mental challenges of survival, keep motivation high, and foster a spirit of hope, even in the most dire circumstances.

20.1 - Understanding the Psychological Toll of a Long-Term Crisis

A survival situation tests not just your physical endurance but also your mental and emotional limits.

The prolonged uncertainty, fear of the unknown, isolation from society, and the pressure of providing for your family all take their toll. Understanding the **psychological challenges** that lie ahead will help you prepare for them, just as you would prepare for physical threats.

Common Psychological Challenges in Long-Term Crises

1. **Stress and Anxiety**: The constant fear of the unknown, the pressure to provide for loved ones, and the insecurity of not knowing how long the crisis will last create a heightened state of stress and anxiety. Over time, chronic stress can lead to physical symptoms such as headaches, fatigue, insomnia, and even heart issues.
2. **Isolation and Loneliness**: Humans are social beings, and long-term isolation or disconnection from friends, family, and society can lead to **loneliness**, **depression**, and **feelings of helplessness**. In a bug-in situation, even within a family unit, the isolation from the outside world can become overwhelming.
3. **Fear and Uncertainty**: One of the most destabilizing aspects of a crisis is the uncertainty of how long it will last, whether external help will come, and if you'll have enough resources to make it through. This constant fear can lead to paralysis, poor decision-making, and irrational thinking.

4. **Burnout and Fatigue**: The daily grind of survival—securing resources, defending your home, and caring for others—can lead to **mental and physical exhaustion**. Over time, this results in burnout, where you become less effective and more prone to mistakes and emotional outbursts.
5. **Hopelessness and Despair**: In extended crises, it's easy to lose hope. If resources run low or the crisis stretches on without any sign of resolution, people can feel trapped or helpless. Maintaining hope in these situations is one of the biggest psychological challenges you'll face.

Anticipating the Psychological Effects

By understanding these common psychological challenges, you can begin to **anticipate** them and prepare your mind to face them head-on. SEALs are trained to expect extreme psychological stress in combat situations, allowing them to mentally prepare for the pressures they'll face. Similarly, by acknowledging that fear, stress, and loneliness are part of the survival experience, you can develop strategies to mitigate their effects.

20.2 - Building Mental Resilience: The Navy SEAL Mindset

Navy SEALs are known for their incredible mental toughness and resilience, traits that are instilled through years of training and discipline. This mental toughness enables them to perform under the most extreme conditions and to push through physical and psychological limits that would break most people. In a survival scenario, adopting a similar mindset will help you maintain focus, stay calm, and keep a sense of control, even when the situation feels overwhelming.

Components of Mental Resilience

1. **Self-Discipline**: Mental resilience begins with self-discipline. SEALs train to maintain strict discipline in every aspect of their lives, from physical fitness to mental focus. In a survival situation, self-discipline means sticking to routines, controlling your emotions, and staying productive, even when motivation is low.
2. **Adaptability**: One of the core principles of SEAL training is **adaptability**—the ability to adjust to rapidly changing situations without becoming overwhelmed. In a survival scenario, this means being able to pivot from your original plans, adjust to new challenges, and embrace the unexpected without losing your focus.
3. **Emotional Regulation**: SEALs learn to **control their emotions**, particularly in high-stress situations. Emotional regulation is the ability to recognize your emotions without letting them take control of your actions. Whether it's fear, anger, frustration, or sadness, you must learn to

manage these emotions in order to make clear-headed decisions.
4. **Positive Visualization**: SEALs use **visualization techniques** to mentally rehearse success before it happens. By visualizing positive outcomes—whether it's surviving a mission or accomplishing a goal—you build confidence in your ability to overcome challenges. In a survival scenario, regularly visualizing positive outcomes can help you stay motivated and optimistic, even when the situation looks bleak.
5. **Purpose and Focus**: SEALs operate with a strong sense of **purpose**, which keeps them focused on the mission, even under extreme stress. In a long-term crisis, maintaining a sense of purpose—whether it's protecting your family, surviving until help arrives, or achieving a specific goal—will help you stay mentally strong.

Mental Toughness Exercises

To build your mental resilience in preparation for a survival scenario, consider practicing **mental toughness exercises**. These can help train your mind to deal with stress, stay focused, and remain calm under pressure.

- **Cold Exposure**: One exercise used by SEALs is **cold exposure**, such as cold showers or ice baths. This forces your mind to deal with

discomfort and teaches you how to remain calm and focused even in physically challenging situations. Over time, you'll become more resilient to stress and better able to handle discomfort.
- **Controlled Breathing**: Another key technique is **controlled breathing**. Deep, rhythmic breathing (such as the **box breathing** technique) calms the nervous system and helps you regain focus when you feel stressed or overwhelmed. Practice deep breathing exercises daily to improve your ability to regulate emotions and stress.

20.3 - Creating Structure and Routine for Psychological Stability

One of the greatest challenges in a long-term survival situation is the breakdown of **daily routines**. Without the normal structure of work, social life, and external obligations, time can feel chaotic, leading to feelings of disorientation and anxiety. SEALs maintain strict routines, even in chaotic environments, because routine provides a sense of order and control.

In a long-term bug-in scenario, creating and maintaining a daily **structure** for yourself and your household is crucial for maintaining psychological stability. Routine not only helps you stay productive but also gives you and your group a sense of **normalcy** in an otherwise unpredictable situation.

Establishing a Daily Routine

A daily routine in a survival scenario should include a balance of productive tasks, rest, and recreation. By setting clear expectations for each day, you'll create a rhythm that helps reduce stress and anxiety.

Productive Tasks

Start each day by outlining a set of productive tasks that contribute to your survival plan. These tasks could include:

- **Food Production**: Gardening, hunting, or foraging.
- **Resource Management**: Inventorying supplies, rationing food, or filtering water.
- **Security**: Checking defenses, rotating watch shifts, or practicing self-defense skills.
- **Maintenance**: Performing necessary repairs or fortifying your shelter.

Having a set of specific tasks each day gives everyone in your household a purpose and reduces the sense of helplessness that can arise in long-term crises.

Physical Activity and Exercise

Physical health is deeply connected to mental well-being. Even in confined spaces or when resources are limited, maintaining a regular **exercise routine** helps reduce stress, improves mood, and keeps your body strong. SEALs often use bodyweight exercises, which require no equipment, to stay fit in any environment.

Create a daily **exercise regimen** that includes:

- **Strength Training**: Push-ups, sit-ups, squats, and lunges.
- **Cardio**: Jumping jacks, burpees, running in place, or walking.
- **Stretching**: Regular stretching or yoga to keep muscles flexible and reduce tension.

Physical activity also serves as a way to release built-up stress and energy, helping to prevent emotional outbursts or frustration within the group.

Rest and Relaxation

In addition to productive tasks and exercise, it's critical to schedule time for **rest and relaxation**. Burnout is a serious risk in long-term survival situations, and maintaining your mental and emotional well-being requires periods of downtime. Make space for recreational activities like reading, playing games, or simply enjoying quiet moments. These activities allow your mind to decompress and recharge.

The Power of Routine in Group Dynamics

If you're leading a household or group, establishing a clear daily routine also helps keep the group organized, motivated, and cohesive. When everyone knows what to expect and has specific tasks to focus on, it reduces confusion, conflict, and stress. SEAL teams operate with precise routines, even in combat situations, because structure is critical to maintaining team cohesion under pressure.

20.4 - Coping with Isolation and Loneliness

One of the most difficult aspects of a long-term bug-in scenario is **isolation**. Without regular social contact, feelings of loneliness can creep in, leading to **depression**, **anxiety**, and a sense of hopelessness. This is especially challenging in situations where access to the outside world is cut off, and people are confined to small spaces for extended periods.

The Effects of Isolation

Isolation can have profound effects on mental health, including:

- **Depression**: Extended periods of loneliness can lead to **depressive symptoms**, such as loss of motivation, difficulty concentrating, and feelings of sadness or worthlessness.
- **Anxiety**: The uncertainty of the situation, combined with a lack of social support, can increase feelings of **anxiety** and **fear**.
- **Cognitive Decline**: Social isolation has been linked to cognitive decline, including **memory**

problems and **slowed thinking**. Without regular mental stimulation or conversation, people can become mentally stagnant.

Combatting Loneliness and Maintaining Social Connections

Even if you're isolated with just a few family members, there are ways to maintain **social connections** and combat loneliness. SEALs who operate in small teams often rely on strong interpersonal bonds to overcome the challenges of isolation during missions.

Strengthen Relationships Within Your Household

If you're bugging in with family members or a small group, make an effort to **strengthen the relationships** within your household. Spend time together, engage in group activities, and check in on each other's emotional well-being. A sense of camaraderie and mutual support can go a long way in reducing feelings of loneliness.

Stay Connected Remotely

If the crisis allows, find ways to stay connected with the outside world, even if only remotely. Use technology (such as radios, phones, or computers) to communicate with extended family, friends, or other survivors. Hearing the voices of loved ones or staying informed about what's happening outside your immediate area can reduce feelings of isolation.

Create New Social Rituals

In survival situations, creating **social rituals** can help maintain a sense of community. Even something as simple as gathering for meals, playing a game, or telling stories in the evening can create moments of connection that help stave off loneliness.

20.5 - Maintaining Hope and Motivation in Long-Term Survival

Perhaps the most difficult aspect of a prolonged crisis is maintaining **hope**. When the days turn into weeks and the weeks into months, it's easy to lose sight of the end goal. SEALs face similar challenges during long deployments, often enduring months of uncertainty and danger. The ability to keep going, even when the outcome is unclear, is what separates those who survive from those who give up.

The Importance of Hope

Hope is not just an abstract concept—it's a **survival tool**. When people lose hope, they stop trying. They become complacent, unmotivated, and eventually give up on themselves and others. As a leader, whether for your family or a larger group, it's your responsibility to keep the flame of hope alive, even in the darkest moments.

Celebrate Small Victories

In a long-term survival scenario, finding reasons to celebrate—even the smallest victories—can make a huge difference in morale. Did you successfully grow your first vegetables? Did you repair a critical piece of equipment? Did your group work together to solve a problem? Celebrate these milestones. Each small victory is a reminder that progress is being made, even if the overall situation is still uncertain.

Set Short-Term Goals

Survival is a marathon, not a sprint. Instead of focusing solely on the ultimate goal of getting through the crisis, break it down into **short-term goals**. These might be daily, weekly, or monthly objectives that give you a sense of forward momentum. Short-term goals keep you focused and give you a sense of accomplishment, even when the larger picture remains uncertain.

Maintain a Positive Outlook

Positive thinking is a powerful tool. While it's important to be realistic about the challenges you face, it's equally important to focus on what you can control and the progress you've made. SEALs train to find the positives in every situation, no matter how dire. By focusing on the things that are going right, you build confidence in your ability to handle whatever comes next.

Inspire Others

As a leader, your attitude directly impacts the morale of those around you. If you project **hope and determination**, others will follow your lead. If you give up, it's likely that others will too. Even in moments of doubt, remind yourself of the responsibility you have to inspire and motivate others to keep pushing forward.

20.6 - Final Thoughts: Surviving and Thriving Through Mental Strength

In a long-term survival scenario, mental and emotional resilience are your most powerful tools. As much as food, water, and shelter are essential, so too is the ability to stay mentally strong, adapt to changing circumstances, and maintain hope. Just as Navy SEALs train their bodies for physical endurance, they train their minds for psychological resilience, knowing that mental strength is often the key to overcoming the toughest challenges.

By applying the principles of **mental toughness**, **emotional regulation**, and **positive thinking**, you can lead yourself and your household through even the most challenging situations. Survival is not just about enduring hardship—it's about finding ways to thrive, even in adversity. With the right mindset, you can overcome the fear, uncertainty, and isolation of a long-term crisis, emerging stronger on the other side.

Remember: You are more resilient than you think, and with preparation, discipline, and hope, you can navigate any crisis.